Guides to Information Sources

Information Sources in

Grey Literature

Guides to Information Sources

A series under the General Editorship of
D. J. Foskett, MA, FLA
and
M. W. Hill, MA, BSc, MRIC

This series was known previously as 'Butterworths Guides to Information Sources'.

Other titles available are:

Information Sources in Metallic Materials
  edited by M. N. Patten

Information Sources in the Earth Sciences (Second edition)
  edited by David N. Wood, Joan E. Hardy and
  Anthony P. Harvey

Information Sources in Polymers and Plastics
  edited by R. T. Adkins

Information Sources in Energy Technology
  edited by L. J. Anthony

Information Sources in the Life Sciences
  edited by H. V. Wyatt

Information Sources in Physics (Second edition)
  edited by Dennis F. Shaw

Information Sources in Law
  edited by R. G. Logan

Information Sources in Management and Business (Second edition)
  edited by K. D. C. Vernon

Information Sources in Politics and Political Science: a survey
worldwide
  edited by Dermot Englefield and Gavin Drewry

Information Sources in Engineering (Second edition)
  edited by L. J. Anthony

Information Sources in Economics (Second edition)
  edited by John Fletcher

Information Sources in the Medical Sciences (Third edition)
  edited by L. T. Morton and S. Godbolt

Information Sources in
Grey Literature
Second Edition

C. P. Auger

BOWKER-SAUR

London • Edinburgh • Munich • New York
Singapore • Sydney • Toronto • Wellington

British Library Cataloguing in Publication Data
Information sources in grey literature. – 2nd ed. – (Guides to information sources).
  1. Research reports
  I. Auger, C. P. (Charles Peter)   II. Use of reports literature
  III. Series
  001.4'33

  ISBN 0–86291–871–5

Library of Congress Cataloging-in-Publication Data
Auger. Charles P. (Charles Peter)
    Information sources in Grey literature / by C.P. Auger.
        p.   cm.   (Guides to information sources)
    Rev. ed. of: Use of reports literature, 1975.
    Includes bibliographical references.
    ISBN 0–86291–871–5 : £30.00
    1. Technical reports. 2. Technical literature. I. Auger,
Charles P. (Charles Peter). Use of reports literature. II. Title.
III. Series: Guides to information sources (London, England)
T10.7.A87   1989
025'.02'0285416—dc20   89–23980
                                                                       CIP

Bowker-Saur Ltd is the library and information division of Butterworths, Borough Green, Sevenoaks, Kent TN15 8PH

Typeset by Dataset Marlborough Design, Oxford
Cover design by Calverts Press
Printed on acid-free paper
Printed and bound in Great Britain by
Biddles Ltd, Guildford and King's Lynn

Series Editors' Foreword

Daniel Bell has made it clear in his book *The Post-Industrial Society* that we now live in an age in which information has succeeded raw materials and energy as the primary commodity. We have also seen in recent years the growth of a new discipline, information science. This is in spite of the fact that skill in acquiring and using information has always been one of the distinguishing features of the educated person. As Dr Johnson observed, 'Knowledge is of two kinds. We know a subject ourselves, or we know where we can find information upon it.'

But a new problem faces the modern educated person. We now have an excess of information, and even an excess of sources of information. This is often called the 'information explosion', though it might be more accurately called the 'publication explosion'. Yet it is of a deeper nature than either. The totality of knowledge itself, let alone of theories and opinions about knowledge, seems to have increased to an unbelievable extent, so that the pieces one seeks in order to solve any problem appear to be but a relatively few small straws in a very large haystack. That analogy, however, implies that we are indeed seeking but a few straws. In fact, when information arrives on our desks, we often find those few straws are actually far too big and far too numerous for one person to grasp and use easily. In the jargon used in the information world, efficient retrieval of relevant information often results in information overkill.

Ever since writing was invented, it has been a common practice for men to record and store information; not only fact and figures, but also theories and opinions. The rate of recording accelerated after the invention of printing and moveable type, not because that in itself could increase the amount of recording but because, by making it easy to publish multiple copies of a document and sell them at a profit, recording and distributing information became very lucrative and hence attractive to more people. On the other hand, men and women in whose lives the discovery of the handling of information plays a large part usually devise ways of getting what they want from other people rather than from books in their efforts to avoid information overkill. Conferences, briefings, committee meetings are one means of this; personal contacts through the 'invisible college' and members of one's club are another. While such people do read, some of them voraciously, the reading of published literature, including in this category newspapers as well as books and journals and even watching television, may provide little more than 10% of the total information that they use.

Computers have increased the opportunities, not merely by acting as more efficient stores and providers of certain kinds of information than libraries, but also by manipulating the data they contain in order to synthesize new information. To give a simple illustration, a computer which holds data on commodity prices in the various trading capitals of the world, and also data on currency exchange rates, can be programmed to indicate comparative costs in different places in one single currency. Computerized data bases, i.e. stores of bibliographic information, are now well established and quite widely available for anyone to use. Also increasing are the number of data banks, i.e. stores of factual information, which are now generally accessible. Anyone who buys a suitable terminal may be able to arrange to draw information directly from these computer systems for their own purposes; the systems are normally linked to the subscriber by means of the telephone network. Equally, an alternative is now being provided by information supply services such as libraries, more and more of which are introducing terminals as part of their regular services.

The number of sources of information on any topic can therefore be very extensive indeed; publications (in the widest sense), people (experts), specialist organizations from research associations to chambers of commerce, and computer stores. The number of channels by which one can have access to these vast collections of information are also very numerous, ranging from professional literature searchers, via computer intermediaries, to Citizens' Advice Bureaux, information marketing services and information brokers.

The aim of the Guides to Information Sources (formerly Butterworths Guides to Information Sources) is to bring all these sources and channels together in a single convenient form and to present a picture of the international scene as it exists in each of the disciplines we plan to cover. Consideration is also being given to volumes that will cover major interdisciplinary areas of what are now sometimes called 'mission-oriented' fields of knowledge. The first stage of the whole project will give greater emphasis to publications and their exploitation, partly because they are so numerous, and partly because more detail is needed to guide them adequately. But it may be that in due course the balance will change, and certainly the balance in each volume will be that which is appropriate to its subject at the time.

This book has its origins in the *Use of Reports Literature*, published in 1975 in the series 'Information Sources for Research and Development'. The present volume incorporates and updates some of the material from the earlier work; it also includes much that is new. After such a long interval, the contributors who could be contacted felt unable to revise their chapters, whilst several had moved to addresses unknown. The entire text therefore has been revised and rewritten by the author alone.

D.J. Foskett
Michael Hill

Preface

Colour coding in literature is not new – for years we have had yellow backs (cheap editions of novels bound in yellow boards and sold on Victorian railway stations); white papers (official documents dating from 1899 and printed on white paper); green papers (discussion documents from the government, provided with green covers, which first appeared in 1967); blue books (official reports which were issued in blue covers); black books (such as the *Black book of the exchequer*); and red data books (various lists of endangered species). To complete the literary spectrum there are of course purple patches and passages, and brown studies.

Recognition of the value of colour coding is not confined to the British Isles. In the United States the term *Blue book* is used to denote various official publications, especially the state manuals which give biographical and other details about government employees. In France the colours are yellow (livre jaune) and, more and more frequently, white (livre blanc). In Germany, too, the government favours white (Weissbuch), whilst in Italy the preference is for green (libro verde), and in Belgium the choice is grey (livre gris).

The latest addition, grey literature, brings a new set of problems, not least because grey has a very negative connotation (dull and dismal grey skies, for example), quite apart from its being the indicator of uncertainty, vagueness and imprecision. On the positive side there is no question that grey literature is proving to be a convenient shorthand term with which to encompass a whole range of difficult-to-define publications not usually available through the normal bookselling channels. There now seems to be no going back, and what used to be known simply as the reports literature has been expanded to include the various categories of documents discussed in the following pages. Indeed the very uncertainty is one of the justifications for the work, because whilst the term grey literature may be finding wider and wider acceptance, it is still little understood. Certainly those writers who create it (engineers, scientists, educationalists, government officials and representatives of various bodies and institutions) rarely show a flicker of recognition whenever the term is mentioned to them, whilst the specialists responsible for its identification and dissemination (librarians, documentalists, publishers and publicists) sometimes profess a nodding acquaintance with the concept, but usually no more.

Fortunately there is a body of experts around the world on whose opinions and writings I have been able to draw, and my thanks go to all the people, too numerous to mention here, who have provided advice and guidance. The views expressed however are solely mine, especially those on how to define the boundaries of grey literature. The expression 'outside the booktrade' is crucial and may need rethinking as the publishing world itself undergoes radical changes. Comments on this, and other aspects of grey literature, will always be most welcome.

Finally it is worth pointing out that there is a considerable resemblance between grey literature and the greyhound – a dog which everyone recognizes by certain well-defined features (it is tall, slender, with great speed and keen eyesight) but which nobody expects to have a coat which is invariably grey.

Contents

Contents

List of Figures

CHAPTER ONE

The nature and development of grey literature

Introduction

Whenever concern is expressed at the continually increasing quantity of grey literature, and at the difficulty it presents to librarians, information workers and documentation experts on the one hand and to readers and users on the other, the reply which is often given is that such publications are not intended to form a part of the permanent literature, and consequently any problems that arise will be purely temporary. Unfortunately this sanguine view is belied by the overwhelming evidence of the durability of grey literature and by its frequent citation in catalogues, bibliographies and reading lists of all kinds.

Those who write and issue grey literature do so because such documents offer a number of advantages over other means of dissemination, including greater speed, greater flexibility and the opportunity to go into considerable detail if necessary. Another important reason why grey literature has attained importance as a separate medium of communication is because of an initial need for security or confidentiality classifications which prevent documents being published in the conventional manner. Over the years grey literature has come to constitute a section of publications ranking in importance with journals, books, serials and specifications, and the time has come to grant it full recognition. Any subjective feelings that grey literature is being used and quoted more and more are borne out by the many independent references to it, and by the emergence of databases devoted specifically to standardizing its identification and to improving its accessibility.

Difficulties arise because whereas most of the different categories of conventional publications are subject to well-established systems of bibliographical control, grey literature still poses awkward questions in terms of identification and availability. Quite often the term 'half published' is used to describe such literature. Precisely because grey literature is so amorphous and intended for a wide variety of purposes, it is not obliged to conform to the standards of presentation imposed by the editors and publishers of conventional publications nor to the rigours of a refereeing system.

A tremendous number of items appearing in the grey literature are initially prepared with a known and limited readership in mind, and often they carry distribution lists as evidence of this. Often, too, copies are numbered individually so that each can be accounted for. So long as items of grey literature remain within their original restricted environment, difficulties of identification and accessibility remain relatively few. Complications begin to occur when details of such publications reach persons other than those originally envisaged.

This state of affairs can come about through a deliberate policy of announcing to anyone interested the news that publications in the grey literature are now available generally, or it may happen incidentally because documents have been quoted openly and sometimes unthinkingly along with other more easily traceable references.

Either way it is at this stage that grey literature undergoes the transition from the category restricted and/or temporary to the category open and permanent, whatever the intentions of the originators. At this point, too, questions of priority arise and whether or not an item of grey literature has a rightful place in the primary sources of a subject can become a matter for dispute when the race is on to be the first into print. Nevertheless once examples of grey literature pass into the conventional literature, they tend to remain there, and are quite frequently quoted and requoted in grey literature form long after they have passed into other, more conventional types of publication.

Immediately grey literature publications become referenced in the open literature, interested would-be readers need to be able to ask for them in the correct manner – no easy task when a document may be characterized by a range of identifying systems such as accession numbers or series numbers, or on the other hand, simply not given any identifiers at all. Equally, the library or other organization asked to supply a document must have ready access to a location. Since in practice grey literature is notified to national copyright and bibliographic agencies in an extremely haphazard manner, the compilation of location indexes and the establishment of comprehensive collections have to rely heavily on co-operative arrangements.

Because grey literature presents problems in identification and acquisition and also because it has developed into a key format in the literature of practically all branches of knowledge, strenuous but varyingly effective attempts have been made to bring the situation under control, and an indication of the degree to which success has been achieved in so far as the reader is able to tap the information sources available by identifying and specifying what he or she wishes to see, and the libraries and other agencies for their part are able to deliver, is the purpose of the present work.

Definitions

The first question to address is what exactly does the grey literature constitute in terms of publications? In an earlier book in the present series, Chillag (1985) gives Wood's widely quoted definition of grey literature as: 'literature which is not readily available through normal bookselling channels, and therefore difficult to identify and obtain'. Examples of grey literature include reports, technical notes and specifications, conference proceedings and preprints, translations, official publications, supplementary publications and data, trade literature, and so on.

It has to be recognized, however, that not all material in all these categories is grey as defined above: conference proceedings may ultimately be published as books or in journals, many official publications become commercially available, and so do, for example, cover-to-cover translated journals.

Poor availability is only one characteristic of grey literature – others include poor bibliographic information and control, non-professional layout and format and low print runs.

Van der Heij (1985) has pointed out that some synonyms for 'grey' used in the literature are 'non-conventional', 'informal', 'informally published', 'fugitive' and even 'invisible'. He also reminds us that documents may be unconventional in many ways, and many conventionally published documents show greyish aspects.

Some writers, e.g. Griffin (1982), have included patents and standards among the constituents of grey literature, but most people would argue that whilst such items may not be readily available through normal bookselling channels (and are not expected to be), they are in fact published through well recognized agencies which observe long established conventions of strict bibliographic control and identification systems. Patents and standards are of course frequently mentioned in publications which announce details of new grey literature documents, but for the purposes of this book they are treated as being outside the accepted definitions, notwithstanding the fact that many British Standards have grey covers!

The problem of trying to define items of grey literature is one which is well known to the life scientist – it is rather like trying to define the species. Everyone can recognize a piece of grey literature when they see it, but it is not easy to write an explanation which can cover all the exceptions.

Another way of arriving at a definition is to examine how a major journal with the avowed intent of publicizing grey literature tackles the problem. The journal in question is the British Library Document Supply Centre's (BLDSC) *British reports translations and theses*, which has three main sections:

(1) Humanities, psychology and social sciences
(2) Biological and medical sciences
(3) Mechanical, industrial, civil and marine engineering

In the case of the biological and medical sciences, and also with the engineering disciplines, the categories of publications will be familiar to workers in the field as items of reports literature under a new name. The big changes have been in the humanities and social sciences, particularly in two specific areas:

(1) Economics and economic history
(2) Urban, regional and transport planning

Thus BLDSC is collecting publications such as market surveys and local development studies, and adding the relevant details to the grey literature data bank. Such local material is also the province of other organizations, notably: the Department of the Environment and the Department of Transport, which jointly maintain collections devoted to structure and local plans, public enquiry documents, and public transport plans; the London Research Centre (formerly the GLC library); and the Planning Exchange in Glasgow.

As the list of categories embraced by the term grey literature gets longer, so the general understanding of the concept grows less. At least scientists and engineers and research workers of all types would show some flicker of recognition when asked their views on the reports literature; now when confronted with the term grey literature, their reaction is one of puzzlement.

Even in the library profession itself, uncertainty about what is meant by grey literature still persists, and only gradually is the term being accepted as a description of what used to be known as 'reports literature', 'unconventional literature' and 'literature not available through the book trade'.

Only time will tell how far the boundaries of grey literature are likely to expand. Certainly two categories of publications spring to mind which possibly did not feature in the idea of grey literature when first conceived, yet which most definitely meet the requirement of not being available through normal bookselling channels.

The first category comprises publications issued by pressure groups and similar bodies with a special point to make. Often such organizations need to publish quickly, their funds are limited, and there is no scope for the niceties of sale or return and trade discounts. Every penny of the cover price is needed by the pressure group, and in consequence sales are achieved by direct mail or specialist outlets. Many pressure groups promulgate their message through a publication or series of publications, make their point (or not), as the case may be, and then cease activity. Others grow in stature, sometimes because they receive support, sometimes because they encounter resistance, and their publications become sufficiently well known and in demand to effect a transition from the grey category to conventional outlets. Examples of pressure groups' publications are reviewed in the national press every week, and examples are quoted elsewhere in this book.

The second category embraces privately published material, which can extend from slim volumes of poetry and detailed family histories to topical stories presented with a particular point of view. The former will probably not end up in grey literature indexes or collections, but the latter may if they attract sufficient notice. An excellent example is the Lonrho book *A Hero from Zero* (Rees-Mogg, 1988), an account of the 1985 Harrods takeover by Mohamed Fayed. Forty thousand copies were printed, but as the *Independent* noted 'it could not be bought in the bookshops'.

No consideration of the definition of grey literature is complete without an examination of the term 'book trade' itself. In Great Britain the dominant characteristic of the book trade is its diversity – it produces as many titles annually as the United States with a turnover and a domestic market only one fifth of the size. Without the protection of the Net Book Agreement (such protection does not exist in the United States), this diversity will be at risk. The Net Book Agreement – which guarantees authors and publishers that books will not be sold below cover prices – was at first just an agreement, but it received the force of law in 1962 when retail price maintenance was generally abolished in Britain. Pressures are mounting to abolish the Net Book Agreement, as the concept of books as just another form of merchandise subject to market forces gathers wider acceptance. The book trade continues to resist 'but in its heart it is fatalistically resigned' (Sutherland, 1988). If abolition does come it will have a profound effect on the notion of grey literature, not least because more and more publications will find themselves outside the book trade.

Development of grey literature

The appearance of the idea of grey literature is not new – Schmidmaier (1986) gives a quotation from the 1920s:

> No librarian who takes his job seriously can today deny that careful attention has also to be paid to the 'little literature' (Kleinschrifttum) and the numerous publications not available in normal bookshops, if one hopes to avoid seriously damaging science by neglecting these.

Schmidmaier gives a further example from the 1930s:

> For years existing systems have been supplying users with several categories of grey literature; example: bibliographic registering of new publications not available in bookshops within the framework of the Deutsche National Bibliographie since 1931.

The current scene with respect to grey literature in Germany has been described by Hasemann (1986).

In the United Kingdom the term grey literature began to acquire general currency in the later 1970s; it received an important endorsement when the considerable problems of acquisition, bibliographic control and access led the Commission of the European Communities to organize in co-operation with the British Library Lending Division (as it was then known) a seminar on grey literature in York in December 1978. The main result of the seminar was to pave the way for the setting up of a grey literature database called SIGLE (*System for information on grey literature in Europe*). The development of SIGLE is discussed in the next chapter.

At this point it is worth trying to answer a frequently asked question: what is the difference between reports literature and grey literature? *Table 1.1* tries to distinguish between the two types.

Still on the subject of distinguishing one category from another, mention must be made of ephemera, which is the collective name given to material which carries a verbal illustrative message and is produced by printing or illustrative processes, but not in the standard book pamphlet or periodical format. Makepeace (1985), in his study on ephemera, notes that 'probably the simplest of the definitions that have been advanced for ephemera are not really definitions as such, but merely alternative names such as 'non-book

TABLE 1.1. Reports literature and grey literature: some key features

	Reports Literature 1940s	*Grey Literature* 1980s
Term first used		
Characteristics	Accounts of projects and investigations sponsored by governments and agencies, with a heavy emphasis on defence contracts.	All types of literature not available through the normal bookselling channels, including reports, trade literature, translations and ad hoc publications.
Bibliographic control	Usually two tier: (1) report code series by originators; (2) accession numbers by issuing agencies.	Haphazard, with some use of ISBNs.
Format	Paper copies and microfiche prevail.	Printed works, typescripts, pamphlets and microforms.
Development and status	Well-established format used and understood by the R&D community.	Cause of concern to IFLAs Universal Availability of Publications (UAP) programme; European initiative in the form of SIGLE (System for Information on Grey Literature in Europe).
Availability	Details contained in comprehensive announcement journals (e.g. STAR).	Uncertain: efforts are being made through national document supply centres.

material', 'fugitive material', 'grey literature' or 'miscellaneous material".

The main type of material that is confused with ephemera is one which is called minor publications or local publications, that is books, pamphlets, newspapers, newssheets or other multipage formats with two important characteristics:

(1) They are produced 'uncommercially', and made available free through casual outlets;
(2) They are produced for limited distribution, e.g. within a society or a local area.

Most items of ephemera are produced for short-term purposes (bus tickets, timetables, posters and so on) and although collectable, form no part of a grey literature database. The local material however can have a long term value, and its inclusion in SIGLE is a matter for careful judgement.

Nature of reports literature

The term grey literature is now used, as noted above, to embrace reports, but long before the term grey literature was coined, the expression reports literature had achieved a wide currency and was well understood even if it was regarded as a difficult area. Indeed because reports have a very special significance in the research and development environment, many workers will readily acknowledge that they understand in a general way what is meant by the reports literature, as will of course workers in the library and documentation fields, but when the question becomes more specific: what is meant by a report? – a number of answers seem equally valid.

On a very formal level a report is a document which gives the results or the progress of a research and/or development investigation. Where appropriate it draws conclusions and makes recommendations, and is initially submitted to the person or body for whom the work was carried out. Commonly a report bears a number which identifies both the report and the issuing organization.

Reports in fact are characteristically the products of organizations, and will vary widely in style and method of publication. As will be seen later they may range from a few pages of technical notes to multi-volume works describing the development of large projects. Many users would agree that a report is incapable of strict definition, particularly when considered in relation to similar publications, such as, for example, conference papers. A factor which complicates the issue is the existence of publications which have all the attributes of reports (issuing organization, author, date, typescript format, card covers, etc.) but which paradoxically bear the specific statement 'this document is not a report'. Such warnings are ignored by the user, who assumes that if a document *looks* like a report then it *is* a report, and may be quoted as such.

Another way of arriving at a definition is to consider whether a report is a particular kind of communication medium quite distinct from all other kinds. The essential feature of a report, and in particular of a scientific or technical report, is that it aims to convey certain specific information to a specific group of readers. A report can be the answer to a question, or a demand from some other person or organization or agency for information. This in itself is nothing special, for questions are being asked and demands being made for information every hour of the working day, and the answers are conveyed by many other written means, in particular by letters, by notes and by memoranda. As a rough indication reports will be preferred when the information is of a certain length in terms of pages, has a useful life of at least a few months' duration and is addressed to a number of readers. Letters, notes and memoranda can of course be long in themselves, but generally they will be informally structured and as likely as not carry no identifying serial number or other means of bibliographical control.

It is useful at this point to digress and see what the dictionary says about the Latin word *reportare*, from which the English word 'report' is derived. Basically the Latin word means to bring back, a concept which raises a number of fundamental points of emphasis for any definition of a report. It implies that a person or a corporate body goes out and gets something it is commissioned to get and carries it back to the commissioning agency. In other words a report is one result of an assignment. In a report the objective is generally more definite and has a more imperative shaping effect than in any other form of exposition. It is just this emphasis on the use of the document that distinguishes a report most strikingly from other types of expository writing: it is prepared for a designated reader or readers who have called for specific information or advice.

How do these definitions relate to the actual output of documents described as scientific, technical, business or economic reports? The contents of such documents will be in accordance with whatever standards are followed by the issuing body and clearly these will exhibit wide variations. Since in general reports are not subject to refereeing, it is no bad thing that some degree of uniformity of content if not quality is imposed by calling for certain items to appear in a report as a matter of course. A step in the right direction has been taken by the British Standards Institution (1986) with the publication of its specification for the presentation of research and development reports. The guidance it offers is sound and logical, and it is to be hoped that reports issuing organizations in the United Kingdom will take heed of its precepts. So far however there are few signs of this happening. A similar publication has been issued by the American National Standards Institute (ANSI) (1974), whilst specialist organizations also provide help, as for example the Defence Research Information Centre's *Formal standards for scientific and technical reports* and the National Aeronautics and Space Administration's *NASA publications manual* (Pirelli, 1982).

In practice, in the major reports series discussed later in this work, thousands of reports follow a standard pattern or some modification of it, and are variously described as studies, notes, evaluations, reviews, state-of-the-art surveys, analyses, and the like. On the other hand, many examples can easily be found where a report so described is really the proceedings of a symposium, a selected annotated bibliography, a standard specification, a handbook, a set of statistical tables, a census return and so on.

In 1967 an American Task Group on the Role of the Technical Report (Federal Council for Science and Technology, 1968) drew up a taxonomy of technical reports which enumerates the general types encountered.

These are:

(1) The individual author's preprint, which may end up as a journal article;
(2) The corporate proposal report, aimed at a prospective customer;
(3) The institutional report, the purpose of which is budget justification and image enhancement;
(4) The contract progress report, the most popular species of technical report in circulation, primarily aimed at the sponsor, but also available to an extensive group of interested persons;
(5) The final report on a technical contract effort, generally the most valuable species in the collection, hallmarked by considerable editorial effort;
(6) The separate topic technical report, very close to the journal article, and the legitimate target of journal editors;
(7) The book in report form, typically a review or state-of-the-art survey;
(8) The committee report, the report series descriptions of which follow widely varying codes.

This mixture of documents truly recognizable as reports with documents less obviously so, presents few problems to the user in the course of consulting a major reports series file, but readers unfamiliar with the wide coverage of reports literature could be excused for looking elsewhere for information on conferences, standards or bibliographies. This uncertainty is of course one of the reasons for the introduction of the term grey literature.

An interesting and thoroughly practical definition of a report emerged during a survey of technical reports in Canadian libraries (Mark, 1970), namely an item issued by an organization usually as an 8½ by 11 inch paper document or as a microfiche or microform, frequently identified by a report number. Several of the surveyed libraries said that they considered a publication a technical report if it has no number but looked similar to reports that did have one. Some libraries also indicated that they included in the reports category miscellaneous technical material too important for a vertical file – greyness indeed!

One result of the imprecision in deciding just what constitutes a report has been that many users of the reports literature regard as reports doctoral dissertations and the preprint series and meetings papers of major American scientific, technical and engineering societies. For the same reason translations are often looked upon as reports, especially when individual papers are allotted report-like numbers. Dissertations, preprints and translations do, however, have sufficient distinguishing characteristics to set them apart from reports, even though they still belong to the greater family of grey literature, and their special features are discussed separately in Chapter 5.

If the contents of reports, so-called, can show such variation (to the extent in fact that one important abstracting service – INIS *Atomindex* – lumps them all together under the heading 'non-conventional literature'), then so can their purpose. Thus, as was noted in the taxonomy above, a report may be a once-for-all account of a specific investigation, a progress report, a summary of a project cumulating a whole series of progress reports or an annual report in the traditional manner. A report's purpose directly affects its availability, and the question of issuing agencies and their caution over security classifications is dealt with in the next chapter.

History of reports literature

Whereas the origins of the grey literature concept can be traced back to the 1970s, the history of reports literature goes back to the beginning of the century, and coincides almost entirely with the development of aeronautics and the aircraft industry, and, although nowadays reports are issued on a tremendously wide range of topics, a significant proportion is still concerned with the area now known as aerospace. The series of reports usually accorded the honour of being first in the United Kingdom is the *R & M (Reports and Memoranda) series* of the Advisory Committee for Aeronautics, subsequently the Aeronautical Research Council, which began appearing in 1909. In the United States the aircraft industry has been represented continuously by the National Advisory Committee for Aeronautics (NACA), now known as the National Aeronautics and Space Administration (NASA), which issued its first report on *The behaviour of aeroplanes in gusts* in 1915. Further information on these important reports issuing agencies is given in Chapter 7.

Some authorities, however, consider that these were preceded by publications which were reports in all but name, notably the *Professional papers* of the United States Geological Survey, which began appearing in 1902, and the *Technologic papers* of the National Bureau of Standards, which were published from 1910 onwards as separate items with individual paging and identified by the letter 'T'. These papers contained the results of investigations into materials and methods of testing and were eventually incorporated (in 1928) in the Bureau of Standards *Journal of research*.

The development of the report as a major means of communication, however, dates back only to about 1941, with the establishment on 28 June of that year of the United States Office of Scientific Research and Development (OSRD), whose task was to serve as a centre for mobilizing the scientific resources of the nation and applying the results of research to national defence.

The basis for the expansion was the realization that the report was the most suitable way of presenting the results of thousands of research projects necessary to promote the war effort. With the cessation of hostilities the OSRD was disbanded, but since there was no respite in research and development activities, and consequently no abatement in the flow of reports, there arose an urgent need for a central agency to maintain the bibliographical control system which OSRD had adopted for the identification of the projects in its charge. The result was that in certain areas special agencies were created specifically to organize and disseminate information, especially reports information, which the stimulus of war had enlarged from a trickle to a flood.

In the United States three distinct but interrelated lines of development can be discerned. Firstly, the Publication Board was established in 1945 (and incidentally gave identity to a set of reports which are still being issued today, the *PB series*). Later the Board was absorbed by the Office of Technical Services (OTS), the agency which also originally had the responsibility in the United States for making available the series of reports prepared by the British and American teams that visited Germany and Japan at the end of and immediately after World War II (see below). Subsequently, in 1964, OTS became the Clearinghouse for Federal Scientific and Technical Information (CFSTI); and then in September 1970, the establishment of the National Technical Information Service (NTIS) was announced, to which the CFSTI was transferred and merged.

Secondly, it proved necessary to set up organizations to process reports which were classified in the military sense, and naturally not available simply for the asking. Two early agencies of this type were the Central Air Documents Office (CADO), which in 1948 evolved out of the Air Documents Research Center and Air Documents Division; and the Navy Research Section (NRS), which arose from a contract in 1946–47 between the Office of Naval Research and the Library of Congress. The missions of both were consolidated into the Armed Services Technical Information Agency (ASTIA), a body established in 1951 and which, incidentally, gave birth to another major reports series which still flourishes today, the *AD series* (from Astia Document).

Astia was under the operational control of the US Air Force until 1963, when its name was changed to the Defense Documentation Center (DDC) and its administration transferred to the Department of Defense (DOD). A further name change occurred when the DDC became the Defense Technical Information Center (DTIC) – see Chapter 12.

Thirdly in the lines of development, World War II saw the start and rapid expansion of the nuclear energy industry, an event which in turn generated whole new series of reports literature. In 1942 a central indexing service was provided for the nuclear activities known under the code name 'Manhattan District' by the Metallurgical Laboratory of the University of Chicago. Later the United States Atomic Energy Commission (USAEC) established a Technical Information Service (TIS), afterwards called the Office of Technical Information (OTI), Division of Technical Information Extension (DTE), and eventually named the Technical Information Center (TIC). Subsequent developments involving the Department of Energy are discussed in Chapter 11.

In the United Kingdom the take-off point in the growth area for reports literature was marked by the appearance in the later 1940s of the CIOS (see below) reports and their successors.

No longer used today, these reports, fully described by Poole (1973) in his account of the Technical Information and Documents Unit (TIDU), were at one time eagerly sought and very much in demand. They were written by investigators who had been selected to examine particular fields of German and Japanese industry. The early investigations were mainly concentrated on seeking intelligence vital for the full prosecution of the war, and were carried out by teams of United States and British Military personnel working under the auspices of the Combined Intelligence Objectives Sub-Committee (CIOS). When CIOS came to an end in 1945, its functions were taken over by the British Intelligence Objectives Sub-Committee (BIOS) for the United Kingdom, and by the Field Information Agency, Technical (FIAT) and its parent body the Joint Intelligence Objectives Agency (JIOA) in the United States. With this division of activities and the end of the fighting the emphasis of the investigations shifted from a purely service angle to one of civilian industrial interest.

All the teams submitted reports on their findings, and in many cases also arranged for the transfer of original documents. The investigations in Germany came to an end in 1947, and the printing and publishing of the reports was finished in 1949.

TIDU was part of the British Board of Trade from 1946 until 1951, when it was absorbed by the Department of Scientific and Industrial Research (DSIR). DSIR itself disappeared with the creation of the Ministry of Technology (Min Tech) and the Department of Education and Science (DES) in 1964. In fact the responsibility for the acquisition and announcement of reports of interest to industry (apart from those concerned with nuclear energy) has over the years rested with a succession of government departments: notably with the Ministry of Supply, transformed in 1959 into the Ministry of Aviation; with the Ministry of Technology, as already noted; and with the Department of Trade and Industry (DTI), established in 1970 and responsible for the Technology Reports Centre until the latter's demise in 1981. The consequences of this are examined in more detail in Chapter 12.

The history of the official responsibility for British nuclear energy reports is equally involved. The Ministry of Supply and the Lord President of the Council were responsible until the creation in 1954 of the United Kingdom Atomic Energy Authority (UKAEA). Since then the Authority has undergone considerable change and reorganization, managing throughout to produce a steady stream of reports; details are given in Chapter 11. Those which are made available to the public are often announced through the medium of Her Majesty's Stationery Office (HMSO), and so present an early example of reports series being subjected to adequate bibliographical control.

In recent years the output of scientific and technical reports in particular, but also of reports of all kinds, has grown apace, the bulk of which is still American, but which has received increasing contributions from other countries, notably France and Germany.

Considerable changes have taken place and indeed are still taking place, particularly in methods of announcement and dissemination. In addition, special collections continue to be established and the pattern of usage becomes more diffuse. Any account of the history of reports literature is therefore bound to be incomplete because the story is still unfolding. In fact the continuing development of reports literature is nowhere more clearly demonstrated than in the frequent changes made to the organizations which are officially responsible for reports. This is particularly so with regard to government departments and agencies, and many such developments are noted elsewhere in this volume.

Bibliography

No introductory chapter on the use of grey literature, especially reports literature, is complete without a mention of some of the basic books and papers on the subject. Many are now considerably dated, but all have a value as general background information. The principal titles concerned are:

Butcher, D. (1983) *Official publications in Britain*. London: Bingley
Fry, B.M. (1953) *Library organisation and management of technical reports literature*. Washington DC: Catholic University of America Press
Hernon, P. and McClure, C.R. (1988) *Public access to government information*, 2nd edn. Norwood N.J: Ablex Publishing Corporation (See especially Chapter 10 – *Technical report literature*, by G.R. Purcell).
Weil, B.H. (1954) Ed. *The technical report: its preparation, processing and use in industry and government*. New York: Reinhold
Woolston, J.E. (1953) American technical reports: their importance and how to obtain them *Journal of Documentation*, **9**, (4) 211–219

References

American National Standards Institute (1974) *Guidelines for the formation and production of scientific and technical reports*. ANSI Z39.18: 1974. For details of the 1987 revision, see: Drener R.A.V. Scientific and technical reports: the American National Standard *Bulletin of the American Society for Information Science*, **13** February/March (35) (1987)

British Standards Institution (1986) *Specification for the presentation of research and development reports.* BS 4811: 1972 (1986). London: BSI

Chillag, J.P. (1985) Grey literature. In *Information sources in physics*, ed. D.F. Shaw, ch. 19. London: Butterworths

Defence Research Information Centre. *Formal standards for scientific and technical reports* DRIC–SPEC–1000 (DRIC–BR–27012).

Federal Council for Science and Technology (1968) *The role of the technical report in scientific and technical communication.* Washington: COSATI PB–180944

Griffin, J. (1982) Industrial organisations as producers and users of non-conventional literature. *IAALD Quarterly Bulletin*, **27** (21)

Haseman, C. (1986) Graue Literatur: Beispiele für Kooperation. *Zeitschrift für Bibliothekswesen und Bibliographie*, **33** (6) 417–427

Makepeace, C.E. (1985) *Ephemera – a book on its collection, conservation and use.* London: Gower

Mark, R. (1970) Technical reports in Canadian university libraries. *ASLA Bulletin*, **34** (2) 47–50

Pirelli, T.E. (1982) *The technological report: an analysis of information design and packaging for an inelastic market.* N82–27187 (NASA–TM–84260)

Poole, L.R. (1973) The Technical Information and Documents Unit. In Staveley, R. Ed. *Government information and the research worker.* London: Library Association

Rees-Mogg, W. (1988) A secret best seller. *Independent*, 8 November

Schmidmaier, D. (1986) Ask no questions and you'll be told no lies: or how we can remove people's fear of grey literature. *Libri*, **36**, (2) 98–112

Sutherland, J. (1988) Are books different? *Times Literary Supplement*, 2 December

Van der Heij, D.G. (1985) Synopsis publishing for improving the accessibility of grey scholarly information. *Journal of Information Science* **11**, 95–107

CHAPTER TWO

Grey literature collections and methods of acquisition

Quantities

The problem of defining grey literature spills over into any attempt to try and estimate the number of items already in existence and the quantities added annually. Even the figures quoted by the various abstracting services directly concerned with a category relatively easy to count, reports, do not always make a meaningful aggregate because many items are announced and abstracted in more than one publication. A very careful and time-consuming analysis would be required to arrive at a total based on these sources. Any estimates which are quoted in respect of the total output in any one year tend to be expressed in very round figures. In the past some attempts have been made to assemble figures relating to reports. For example, the famous Weinberg report (President's Science Advisory Committee, 1963) gave a figure of 100 000 government reports issued each year in the United States, and this total has often been quoted. Confirmation of this vast quantity can be obtained from the study on the scientific and technical information explosion by Emrich (1970), which gave the following data:

> Semiformal media (i.e. preprints, technical reports and memoranda) announced in 1967: *US Government research and development reports* 45 000 items; *Scientific and technical aerospace reports* 31 000 items.

A further figure is obtained from *Defense research and development of the 1960s* (Defense Documentation Center, 1970), a cumulation of citations and indexes to over 450 000 reports accessioned to the Defense Documentaton Center and announced in the decade 1960–1969 in the *Technical abstracts bulletin* and *US Government research and development reports*.

Twenty years on, the figures are: *Government reports announcements and index* 59 735 items; *Scientific and technical aerospace reports* 20 247 items.

The introduction to the second edition of the *Dictionary of reports series codes* (Godfrey and Redman, 1973) compiled by the Special Libraries Association notes that at the time of the appearance of the first edition (1962), estimates of the number of reports issued annually varied from 50 000 to 150 000. Since then guesses have ranged much higher but it is very difficult to ascertain a reliable figure. Moreover, if grey literature as a whole is considered, the problem, as Chillag (1982) has noted, becomes one of 'how long is a piece of string'. He suggested that 'depending on definition parameters, in a worldwide all subject context, it is quite reasonable to talk of an output of 100 000–200 000 such items a year'. Certainly the number of bodies issuing reports has grown; the third edition of the *Report series codes dictionary* (Aronson, 1986) records over 20 000 report series codes by nearly 10 000 corporate authors.

The difficulties involved in obtaining a realistic figure for reports issued from a simple count of report numbers have been demonstrated by a number of studies – see for example the comments by Hardick (1987), who cites earlier studies and gives some five examples of her own.

More detailed figures of reports made available in given subject areas are examined later in this work.

Availability and security restrictions

In general grey literature suffers from availability problems because it is often difficult to find out just what is available, but for many categories there are rarely any security restrictions. This however is not the case with reports literature. There can hardly be a more frustrating experience, in view of the wealth of material so adequately documented, than having a request for a report, sometimes after considerable effort and trouble to identify it properly, returned to the applicant marked with the words 'not available'. Sometimes a reason is added, such as 'limited distribution' or 'authorized applicants only'; sometimes no explanation at all is provided. In either case the would-be reader has to face the fact that he has reached the end of the line as far as his application is concerned.

On the face of things, the availability of reports should present few problems. Special organizations have been set up in Great Britain, in the United States, in Germany and elsewhere whose prime purpose is to deal with requests for reports literature. These organizations normally provide a speedy service extremely efficiently, so that when failures do occur the applicant is genuinely baffled.

The reasons for non-availability are legion, and part of the problem may simply be incomplete or incorrect identification – many a reject becomes easy to obtain once a check is made to rectify a vital omission such as an accession number. These factors of course influence success with grey literature as a whole.

More serious difficulties arise when non-availability is due to security classifications imposed by government departments. Typical categories affecting reports are 'secret', 'restricted', 'confidential', 'not for publication' and 'proprietary'. The actual degree of protection which attaches to each security marking will vary from organization to organization. In the context of reports literature, security generally means the safeguarding and protection of classified documents against unlawful dissemination, duplication or observation because of their importance to national defence or security or to the continuing competitive advantage of commercial organizations.

The term 'classified' refers to the degree of secrecy which prevents disclosure to unauthorized persons. Each document is security–classified individually, and is subject to regular review with regard to possible change or declassification. Usually the most stringent classification systems apply to issuing departments, but equally organizations and firms entitled to receive classified documents, in order for example to fulfil official contract obligations, must observe the same strict security measures on their own premises, and regular checks are made to see that they do so.

In general, before classified information can be made available, the intended recipient must be security-cleared. Furthermore the facility where the classified material will be received, stored and used must also be cleared, and there must be a clearly demonstrated 'need to know' for the information in connection with the work being performed. Advice on the regulations governing security–classified reports is best sought by direct contact with the appropriate issuing organizations.

In the United Kingdom the security of government reports is ultimately covered by the Official Secrets Acts 1911, 1920 and 1939. The origins of this legislation have been vividly described by

Cook (1985), who echoes a widely-held view that Section II has aroused the most controversy. In fact in 1972 the Franks Committee reported that Section II should be repealed and replaced by an Official Information Act. According to the Franks Report there are 2000 differently worded charges which can be brought under Section II, but the two basic criminal offences under it are the imparting of unauthorized information, and its receipt. Secrets are classified under four headings:

(1) Top secret;
(2) Secret (i.e. information and material whose unauthorized disclosure would cause serious injury to the interests of the nation);
(3) Confidential;
(4) Restricted.

The 1988 White Paper outlining changes to Section II has been claimed by its critics as being concerned with how to control information more efficiently rather than how to provide the public with more information. Such matters have a direct relevance to grey literature, especially as one of the words used to describe the publications it embraces is fugitive.

The Official Secrets Bill to replace Section II of the Official Secrets Act, 1911, was published in December 1988.

Apart from formal classification under the Official Secrets Act, the British government can resort to other measures which make it difficult to get hold of reports, whose contents have been widely publicized. It can for example restrict access by depositing a copy in the House of Commons Library (as happened with the report on the salmonella epidemic drawn up jointly by the Ministry of Agriculture, the Department of Health and the British Egg Industry Council in 1988); it can restrict copies to 'essential people' (as happened with the Department of Transport's report on the loss of the MV *Derbyshire*).

In the United States a somewhat different situation obtains, because after the Watergate affair and the resignation of President Nixon in 1974, several important acts were passed. The 1966 Freedom of Information Act was already in existence, and it had established the principle that people had a right to know. No longer was it up to them to prove why they wanted information – it was up to the government to prove why it could not be given. The 1974 Privacy Act opened up individuals' files to enable people to inspect and correct; it also blocked access by others.

The 1976 Freedom of Information Act proved even stronger, because under it 'any person may request information held by any executive department government or corporation, government controlled corporation or other establishment in the executive branch of government, including the Executive Office of the President or any independent regulatory agency'. The bodies caught by the Act included the Central Intelligence Agency, the Federal Bureau of Investigation, Treasury, Food and Drug Administration, Internal Revenue, and so on. No reason for an enquiry for information need be given unless it directly affects another person, when a 'balance of interest' principle will be taken into account. There are inevitably some exemptions, and if a department claims information is withheld for defence or foreign relations reasons, a court considering the case must be satisfied that such information is properly classified to prevent damage to those interests.

The 1976 Act is believed to have worked something of a revolution – it has made the Federal government move (especially in the area of consumer issues) and it has exposed government wrong-doing. It has certainly encouraged the investigative journalist, who now has greater access to the grey literature. For example a piece in the British press (May, 1988) about biological warfare quoted a US Army technical report *Biological vulnerability assessment: the US west coast and Hawaii* written in 1986 by William Rose and Bruce Grim. The correspondent points out that the report was classified as secret until he managed to get sections of it released under the Freedom of Information Act.

The Fund for Open Information and Accountability, New York, publishes detailed instructions (termed a 'kit') on how to make a request under the Freedom of Information Act, and provides guidance on monitoring progress, making sure of one's entitlement to material, checking for completeness and the lodging of an administrative appeal.

The administrators of security classified collections are constantly striving to eliminate those documents in their charge which no longer require security protection, largely because the cost figures for maintaining inventories of classified information and the associated time-consuming issue and return procedures are very high. Regrettably classification of one sort or another will always be a characteristic of certain types of reports literature and all the reader not entitled to security clearance can do is to exercise patience, for the information sought may be released one day.

Issuing Agencies

The term 'issuing agencies' originally referred to the various bodies initially responsible for the preparation and distribution of reports. Their very multiplicity and the fact that they did not use normal commercial publishing channels to make the output available was one of the impelling reasons for the establishment of secondary centres, specially designed to handle reports from all quarters. The picture has changed with the recognition that unpublished or hard to obtain material is not confined to reports, which are simply one component in the larger area now known as the grey literature.

In the United Kingdom bodies which typically issue grey literature include:

Associations	Libraries
Churches	Museums
County councils	Private publishers
Educational establishments	Research establishments
Federations	Societies
Institutes	Trade unions
Institutions	Trusts
Laboratories	Universities

Such bodies are invited to contribute to the collection which is being built up at Boston Spa.

In many cases however there is no need to use a secondary distribution centre like the British Library to obtain items of British grey literature: it is simpler to apply direct to the original issuing agency or publishing organization. Some bodies issue regular lists of new publications and invite direct applications for copies. Others publicize their offerings in catalogues and indexes prepared by a variety of organizations willing to act as selling agents and distribution points.

As to the United States the most fruitful procedure for a person seeking details about American grey literature, especially reports, is to try some of the guides and abstracting publications noted below and elsewhere in the book.

A body which has had a great influence on the organization and availability of reports literature for many years and whose principles are still current today is the Committee on Scientific and Technical Information (COSATI).

COSATI, whose governance was assigned to the National Science Foundation in 1971, but is now defunct, was comprised of representatives of the Federal agencies responsible for operating scientific and technical information systems. Its structure and workings were made public through the COSATI annual reports, and its aim was the orderly and co-ordinated development of Federal agency information programmes in the public interest.

A major COSATI contribution to the organization of reports literature is the *Subject category lists*, a uniform subject arrangement for (1) the announcement and distribution of scientific reports which are being issued or sponsored by Executive Branch agencies, and (2) management reporting. The list is a schedule consisting of major fields with a further subdivision of the fields into groups. Scope notes are included for each group. The list has enabled abstracts, citations and the like to be gathered into broad fields or groups for display to the user, and also for distribution purposes.

The COSATI category list (Committee on Scientific and Technical Information, 1964) (*Figure 2.1*) has been widely adopted throughout the world and is the basis of arrangement for a number of abstracting and indexing services in the grey literature.

National approaches

Since so much national effort has gone into providing and administering grey literature, it is not surprising that national collections and specialized announcement journals have been formed to ensure its full and economic exploitation.

How far they have succeeded or are succeeding is not easy to establish. A study by Klempner (1968) investigated the distribution pattern in the United States for the abstracting and indexing services of the Department of Defense, the National Aeronautics and Space Administration, and the Department of Commerce: the results revealed a distribution reaching 27 per cent of educational and non-profit research centres, 31 per cent of industrial research laboratories, and less than one per cent of US manufacturing establishments – a situation regarded as being far from adequate.

Moore and others (1971) studied the difficulties encountered by the users of the National Technical Information Service's indexing and abstracting facilities, and found the main problems were:

(1) Coverage too broad for a single index;
(2) Duplication of material indexed or distributed elsewhere;
(3) Multiple number series;
(4) Inconsistency of bibliographic entries;
(5) Coverage unpredictable.

As if this were not enough, a study conducted more than a decade later by McClure and co-workers (1986) into the effectiveness of a sample of United States academic and public libraries as linking agents, or intermediaries, in the provision of NTIS information services and products revealed a number of barriers which were

NTIS SUBJECT CATEGORY AND SUBCATEGORY STRUCTURE

You can use the Edge Index on the outside back cover to locate the beginning of each category within the Reports Announcements section.

Category 1. Aeronautics

Subcategories: Aerodynamics; Aeronautics Aircraft; Aircraft Flight Control and Instrumentation; Air Facilities.

Category 2. Agriculture

Subcategories: Agricultural Chemistry; Agricultural Economics; Agricultural Engineering; Agronomy and Horticulture; Animal Husbandry; Forestry.

Category 3. Astronomy and Astrophysics

Subcategories: Astronomy; Astrophysics; Celestial Mechanics.

Category 4. Atmospheric Sciences

Subcategories: Atmospheric Physics; Meteorology.

Category 5. Behavioral and Social Sciences

Subcategories: Administration and Management; Documentation and Information Technology; Economics; History, Law and Political Science; Human Factors Engineering; Humanities; Linguistics; Man-machine Relations; Personnel Selection, Training and Evaluation; Psychology (Individual and Group Behavior); Sociology.

Category 6. Biological and Medical Sciences

Subcategories: Biochemistry; Bioengineering; Biology; Bionics; Clinical Medicine; Environmental Biology; Escape, Rescue, and Survival; Food; Hygiene and Sanitation; Industrial (Occupational) Medicine; Life Support; Medical and Hospital Equipment; Microbiology; Personnel Selection and Maintenance (Medical); Pharmacology; Physiology; Protective Equipment; Radiobiology; Strett Physiology; Toxicology; Weapon Effects.

Category 7. Chemistry

Subcategories: Chemical Engineering; Inorganic Chemistry; Organic Chemistry; Physical Chemistry, Radio and Radiation Chemistry.

Category 8. Earth Sciences and Oceanography

Subcategories: Biological Oceanography; Catography; Dynamic Oceanography; Geochemistry; Geodesy; Geography; Geology and Mineralogy; Hydrology and Limnology; Mining Engineering; Physical Oceanography; Seismology; Snow, Ice, and Permafrost; Soil Mechanics; Terrestrial Magnetism.

Category 9. Electronics and Electrical Engineering

Subcategories: Components; Computer; Electronic and Electrical Engineering; Information Theory; Subsystems; and Telemetry.

Category 10. Energy Conversion (Non-propulsive)

Subcategories: Conversion Techniques; Power Sources; Energy Storage.

Category 11. Materials

Subcategories: Adhesives and Seals; Ceramics, Refractories, and Glasses; Coatings, Colorants, and Finishes; Composite Materials; Fibers and Textiles; Metallurgy and Metallography; Miscellaneous Materials; Oils, Lubricants, and Hydraulic Fluids; Plastics; Rubbers; Solvents, Cleaners, and Abrasives; Wood and Paper Products.

Category 12. Mathematical Sciences

Subcategories: Mathematics and Statistics; Operations Research.

Category 13. Mechanical, Industrial, Civil, and Marine Engineering

Subcategories: Air Conditioning, Heating, Lighting, and Ventilation; Civil Engineering; Construction Equipment, Materials, and Supplies; Containers and Packaging; Couplings, Fittings, Fasteners, and Joints; Ground Transporation Equipment; Hydraulic and Pneumatic Equipment; Industrial Processes; Machinery and Tools; Marine Engineering; Pumps, Filters, Pipes, Fittings, Tubing, and Valves; Safety Engineering; Structural Engineering.

Category 14. Methods and Equipment

Subcategories: Cost Effectiveness; Laboratories, Test Facilities, and Test Equipment; Recording Devices; Reliability; Reprography.

Category 15. Military Sciences

Subcategories: Antisubmarine Warfare; Chemical, Biological, and Radiological Warfare; Defense; Intelligence; Logistics; Nuclear Warfare; Operations, Strategy, and Tactics.

Category 16. Missile Technology

Subcategories: Missile Launching and Ground Support; Missile Trajectories; Missile Warheads and Fuses; Missiles

Category 17. Navigation, Communications, Detection, and Countermeasures

Subcategories: Accoustic Detection; Communications; Direction Finding; Electromagnetic and Acoustic Countermeasures; Infrared and Ultraviolet Detection; Magnetic Detection; Navigation and Guidance; Optical Detection; Radar Detection; Seismic Detection.

Category 18. Nuclear Science and Technology

Subcategories: Fusion Devices (Thermonuclear); Isotopes; Nuclear Explosions; Nuclear Instrumentation; Nuclear Power Plants; Radiation Shielding and Protection; Radioactive Wastes and Fission Products; Radioactivity; Reactor Engineering and Operation; Reactor Materials; Reactor Physics; Reactors (Power); Reactors (Non-power); SNAP Technology.

Category 19. Ordnance

Subcategories: Ammunition, Explosives, and Pyrotechnics; Bombs; Combat Vehicles; Explosions, Ballistics, and Armor; Fire Control and Bombing Systems; Guns; Rockets; Underwater Ordnance.

Category 20. Physics

Subcategories: Acoustics; Crystallography; Electricity and Magnetism; Fluid Mechanics; Masers and Lasers; Optics; Particle Accelerators; Particle Physics; Plasma Physics; Quantum Theory; Solid Mechanics; Solid-state Physics; Thermodynamics; Wave Propagation.

Category 21. Propulsion and Fuels

Subcategories: Air-breathing Engines; Combustion and Ignition; Electric Propulsion; Fuels; Jet and Gas Turbine Engines; Nuclear Propulsion; Reciprocating Engines; Rocket Motors and Engines; Rocket Propellants.

Category 22: Space Technology

Subcategories: Astronautics; Spacecraft; Spacecraft Trajectories and Reentry; Spacecraft Launch Vehicles and Ground Support.

This categorization scheme is the one endorsed in 1964 by the Committee on Scientific & Technical Information (Cosati) of the Federal Council for Science & Technology. A booklet describing these categories is available from NTIS: the NTIS order number is AD–621 200, the price codes are PC A04/MF A01.

Figure 2.1 **The COSATI subject scheme** (with acknowledgements to NTIS)

identified as limiting academic and public library staff and their clientele in obtaining effective access to NTIS and its resources. Some of the more important barriers include:

(1) Library staff have low awareness of specific NTIS information services and products;
(2) Library staff have limited reference knowledge and skills related to specific NTIS information services and products;
(3) Academic and public libraries offer minimal physical access to NTIS materials;
(4) Library staff and selected clientele perceive the costs of NTIS information services and products as excessive;
(5) NTIS has unclear marketing objectives and delivery strategies for the provision of its services and products to academic and public libraries.

No doubt similar barriers are perceived by academic and public libraries in the United Kingdom.

A review of the role of NTIS came from the NTIS itself in the form of a paper by Caponio and Bracken (1987) on the role of the report in technological innovation. In it the authors identified a particular type of user of the NTIS services, the technologist, or research and development practitioner, whose information needs differ widely from the basic or 'frontier' research worker. They observe that the technology transfer process always starts with at least a minimal amount of pertinent information. The technologist has to learn that a desirable technology exists, that information on it is available in accessible formats, and that there is a procedure for acquiring, evaluating and possibly using it. Such technology transfer has been accomplished in large measure by the technical report, and of course for more than forty years NTIS has served as the primary source for the collection and distribution of government-sponsored research and engineering reports, and other documents.

All this serves to highlight the validity of the oft-made statement that grey literature is a difficult area, not least because of identifying who the real users are or should be.

In the United Kingdom the largest collection of grey literature is to be found at Boston Spa, the British Library Document Supply Centre. Stocks held and records maintained are shown in *Figure 2.2* for the year 1988.

The store houses about 1700 miles of film and some seven million individual fiche, expanding at the rate of around 250 000 fiche per year and about 25 miles of roll film. There is about twenty years' expansion space left in the store.

STOCK

Category		Holdings	Annual intake
Reports in microform		3 010 000	130 000
Other reports		325 000	25 000
Dissertations (US)		416 000	6000
Doctoral theses		76 000	6000
Conference proceedings		240 000	20 000
Translations	over	500 000	12 000
Local authority material		22 500	400

RECORDS

Category	Number	Annual additions
British reports translations and	140 000	28 000
theses		
Translations index	573 300	8000
Government publications	339 303	8268

Figure 2.2 **BLDSC grey literature stocks and records (1988 data)**

In terms of document format, the BLDSC reports and microform store consists of:

(1) 60 per cent reports, from agencies such as NTIS, NASA, USDOE, INIS, ERIC;
(2) 20 per cent dissertations and doctoral theses;
(3) 20 per cent 'anything published in microform', serials, monographs, Slavonic material, music, other theses, translations, conferences and large research collections.

Many other countries have organizations charged with arrangements for handling and recording the grey literature. In Europe the list comprises the centres contributing to or associated with the SIGLE database (see below). Material emanating from Japan can be accessed through the Japanese Information Service in Science, Technology and Commerce, the youngest of the specialized information services set up within what is now the Science and Technology Division of the British Library. The service was set up in 1985 and offers access to:

(1) Online searching, including Japanese databases;
(2) Over 3500 Japanese scientific, technological and commercial journals;
(3) Over eight million Japanese patents;
(4) Market and industry surveys, company business and trade information;
(5) Conferences and reports;
(6) Translations, journals and indexes.

In Japan itself there are two central organizations for information activities, namely the National Centre for Science Information Systems (NACSIS), operated under the Ministry of Education and Science; and the Japan Information Centre of Science and Technology (JICST). NACSIS is primarily academic, but will handle theses requests, whilst JICST is the main organization for grey literature. Today JICST collects about 14 000 journals, of which around half are from abroad; it also collects technical reports and conference proceedings. JICST offers an on-line service for bibliographic databases through JOIS (JICST on-line Information System). The current state of grey literature in Japan is summarized in a report based on a field study conducted in Japan by Morita (1988).

In Australia the Commonwealth Scientific and Industrial Research Organisation (CSIRO Australia) is a large and diverse government body wherein each division and institute publishes a regular (annual or biennial) report and lists of its own scientific papers and technical reports. Such documents form the bulk of CSIROs publications, but the Organisation also publishes three corporate services:

(1) ECOS, on science and the environment;
(2) Rural research;
(3) Industrial research news.

In France, the Centre National de la Recherche Scientifique has created a new body, the Institut de l'Information Scientifique et Technique (INIST) to replace the Centre de Documentation Scientifique et Technique (CDST) and the Centre de Documentation Sciences Humaines (CDSH). INIST is intended to be a unique organization exploiting one of the largest specialized collections of information in the world and providing access to the information through a wide variety of products. Special attention will be paid to grey literature, and INIST will be active in the production of the SIGLE database. Current holdings at INIST include:

(1) 30 000 Scientific reports (annual increase 3500);
(2) 40 000 Conference proceedings (annual increase 2000);
(3) 100 000 French theses (annual increase 6000).

In Germany the Central Special Library for technology and the pure sciences, chemistry mathematics and physics consists of two institutions: the library of the University of Hanover (UB) and the Technical Information Library (TIB), also at Hanover. The UB/TIB places special emphasis on grey literature, including unpublished

German research reports, foreign reports and conference proceedings, and publications in East European and East Asian languages and in other languages less familiar in Germany.

The collection at the UB/TIB includes:

(1) 3.1 million books and microforms;
(2) 22 800 journal subscriptions;
(3) 718 000 US dissertations and reports;
(4) 157 000 German dissertations;
(5) 59 000 unpublished German reports;
(6) 24 000 unpublished translations from oriental languages;
(7) 70 000 unpublished Soviet documents;
(8) 160 000 volumes of conference proceedings;
(9) 120 000 papers, preprints and US technical standards.

Canada's national science and technology institution, the National Research Council of Canada (NRCC) was established in 1916 to promote scientific and industrial research. In particular it operates the Canada Institute for Scientific and Technical Information (CISTI), providing a national information network for scientific and industrial researchers. The Institute was created in 1974 and is responsible for building and maintaining the national collection of scientific technical and medical literature. In addition the Institute operates a wide range of services, including document delivery, online information retrieval systems, reference and referral services, and publications. The collection is international in scope and presently consists of about three million titles, including books, conference proceedings, journals, technical reports, theses and reference works. CISTI holds more than two million technical reports on microfiche.

As might be expected, reports and other publications issued as the result of work commissioned or supported by governments and government bodies are announced in the regular publications which give news of government publications in general. This is true in both the United Kingdom and the United States, but in each case only the minimum of details is given (no abstracts or detailed indexes).

Her Majesty's Stationery Office (HMSO), the British Government publisher, issues daily lists and a monthly catalogue with annual cumulations, which in particular announce number of series of reports, including those from the United Kingdom Atomic Energy Authority. Sectional lists are available too, arranged according to the government departments sponsoring the publications listed, as for example Sectional list no 2: *Education*.

The *Monthly catalog of US government publications* is the most comprehensive general index of federal government publications, and it announces numerous technical reports whether they are for sale from the Government Printing Office or not.

The story of government publishing is a subject in itself – on the development of HMSO and USGPO see for example the paper by Rogers (1987). So long as government publications are available through the government publisher, and hence through the book trade, they cannot be regarded as part of the grey literature. However a practice which has been growing is for individual government departments and agencies not to use HMSO and instead issue publications direct, frequently by announcement and review in the general press. At present over fifty per cent of all British official publications are now not published by HMSO, amounting to around 10 000 publications a year by over 400 different organizations, a very significant quantity of grey literature. The reasons for this change may in part be economic, for in 1980 HMSO began operating as a Trading Fund, and in 1988 became an independent agency responsible to the Treasury, with the freedom to compete with the private sector.

The means for keeping track of material not handled by the government publisher is Chadwyck-Healey's *Catalogue of British official publications not published by HMSO*. Annual cumulations go back to 1980 and the current services include:

(1) Bi-monthly issues;
(2) Subject and author indexes;
(3) Periodicals lists;
(4) Indexes to sources (a useful list of 1500 names and addresses);
(5) Document delivery;
(6) Keyword indexes.

A welcome amalgamation of the indexes to HMSO and non-HMSO publications is *Catalogue of United Kingdom official publications* (UKOP), published on CD-ROM by HMSO Books and Chadwyck-Healey. UKOP constitutes a record of UK official publications and the publications of twelve major international organizations (such as the Council of Europe, the International Monetary Fund and the World Health Organisation) from 1980 to the present, and is thus a blend of conventionally published and grey literature.

Acquisition of grey literature

Grey literature can be obtained on a routine basis by a variety of methods, but two of the most commonly used are: (1) exchange agreements with other organizations, if permitted; (2) purchases by subscription or on a single item per order basis.

Various facilities are available – for example deposit accounts and prepaid coupons for NTIS services; monthly standing orders for microfiche copies of certain categories of reports, Unesco book coupons; annual subscriptions to the outputs of various agencies, either in whole or selectively; deposit accounts or standing orders for documents announced through HMSO and special arrangements with booksellers and library supply companies.

Major institutions able to afford and justify subscriptions to entire series of publications clearly enjoy the advantages of an automatic delivery service and also completeness of coverage. Organizations and individuals wishing to acquire or purchase single copies of documents, especially small value items may encounter some difficulties, and it is vital to comply with the 'how to order' instructions which many issuing agencies have drawn up. It is rare for example for the normal book trade channels to be interested in obtaining reports; it is far quicker and more effective to apply direct, and every effort should be made to observe the various procedures specifically designed to speed the handling of orders, such as prepaid coupons and deposit accounts.

In the case of reports care should be exercised when ordering by accession number only, because some items may be more readily available in another form. Thus a request for an item identified by a NASA accession number may turn out to be a patent specification more readily obtained from the regular source for patents.

The availability of grey literature in general is patchy simply because one of the definitions of grey literature is 'items not available through the book-trade'. How the book trade itself is defined in these circumstances is never made clear, except that it generally includes publishers, booksellers and subscription agents. Many items of grey literature can be obtained through the book trade, and booksellers often go to great lengths to fulfil orders. Some include the category 'grey literature' in their catalogues of new and forthcoming publications. This is particularly true of companies in the book trade specializing in library supply activities. On the other hand publishers who issue titles with a limited or local interest insist on direct orders, presumably to avoid giving a trade discount.

The criteria by which grey literature is selected and acquired will depend on a collecting organization's policies or an individual's specific needs. A major constraint will be the size of the purchasing budget.

Another way of acquiring grey literature is by borrowing, and the following remarks apply mainly to procedures operating in Great Britain. At the risk of labouring the point, the first essential to observe is that any request should specifiy the correct bibliographical details, in the case of reports the relevant accession number, in the case of other documents at least the ISBN where one has been allocated. Many a request has been slow in producing results because although the reader has taken great pains to indicate the originating body, the author, the title, the date and other pieces of information, the required accession or order number has been omitted. Many items of grey literature have a whole range of identifiers, and by selecting a few, such as those mentioned above, the reader expects his request to be met, especially when compared with the relatively laconic descriptions needed for books and formal articles. Librarians and information workers are renowned for their detective-like qualities in tracing incomplete references, but eventually their patience wears thin.

The British Library Document Supply Centre has made great strides in streamlining its lending service for grey literature, partly by making its collection as comprehensive as possible and partly by rationalizing its request processing routines. In the case of reports instances sometimes occur when specific items are unexpectedly not held, and there can be delays of up to several weeks whilst attempts are made to obtain them. The British Library has also been obliged to point out that even though reports originate from or are disseminated by US Government agencies, they are not available unless listed in one of the major reports announcement journals and/or released by NTIS or the originating body. Reports so excluded are best sought from the specialist organizations and libraries in the subject areas covering the topics in question. NTIS has a special arrangement when it comes to selling its publications; it uses appointed agents for countries around the world, and the mechanics of the system are discussed in Chapter 12.

Large organizations handling hundreds of thousands of documents each year aim to provide satisfactory services to their users, but unfortunately they sometimes become victims of their own size, and so do not always succeed.

Databases

Until the arrival of SIGLE (*System for information on grey literature in Europe*), databases covering grey literature were developed to provide for specific subject areas such as energy or aerospace, and they made a feature of reports literature in their records. SIGLE is different in that it treats a body of information by form rather than by subject content. Parallels do spring to mind of course, notably in patents (the Derwent indexes) and in standards (BSI Standardline) but SIGLEs broadness of coverage, based on the COSATI subject category scheme once used by the US National Technical Information Service, gives it a very special character. The database is produced by a consortium of European documentation centres with the help of the Commission of the European Communities. SIGLE became operational in 1981: up to 1983 only the UK documentation centre (the British Library Document Supply Centre) input documents outside the fields of science and technology, but from 1983 onwards, there is coverage in all subject fields. Theses have been included only from 1983. At present SIGLE is available through the hosts BLAISE and INKA, and once documents sought have been identified the best route to obtain them is via the appropriate national authority or associated institution. There are national authorities in:

(1)	Belgium	— Association Laboratoire Belge des Industries Electrique et Université Catholique de Louvain;
(2)	Germany	— Fachinformationszentrum Energie, Physik, Mathematik GmbH (FIZ) and Universitätsbibliothek und Technische Informationsbibliothek, Hannover (UB/TIB);
(3)	France	— Centre Nationale de la Recherche Scientifique: Institut de l'Information Scientifique et Technique (successor to CDST and CDSH);
(4)	Great Britain	— British Library Document Supply Centre (BLDSC);
(5)	Ireland	— Institute for Industrial Research and Standards;
(6)	Luxemburg	— Bibliothèque Nationale
(7)	Italy	— Consiglio Nazionale delle Ricerche, Rome;
(8)	Netherlands	— Koninklijke Bibliotheek, The Hague;
(9)	Sweden	— Studsvik Library, Nykoping;
(10)	European Communities	— Office for Official Publications of the EC, Luxemburg.

Ever since SIGLE began the United Kingdom has been by far the biggest contributor to the database, and in 1988 the position was:

(1) United Kindom   38 per cent;
(2) Germany   29 per cent;
(3) France   15 per cent;
(4) Others   balance.

The marketing and promotion of SIGLE are handled by the European Association for Grey Literature Exploitation (EAGLE).

The national contributions to the input are merged into a common database at the Commissariat à l'Energie Atomique – Centre d'Etudes Nucléaires (CEN), Saclay, France. The records in SIGLE aim mainly at identification and document delivery – they are not full catalogue records and they do not contain abstracts.

Two other major European databases concerned with grey literature are the *Conference papers index* and FBR. The former is used for tracing conferences from a small amount of information, finding out if they are held at the British Library Document Supply Centre (where the data is compiled), and requesting loans; the latter is a record of the literature on projects sponsored by the BMFT (Bundesministerium für Forschung und Technologie), the Federal ministry for research and technology. Both databases have hard copy versions, the former published by Boston Spa, and the latter called *Forschungsberichte aus Technik und Naturwissenschaften (Reports in the fields of science and technology published in the Federal Republic of Germany). Forschungsberichte* is a co-operative editorial activity between the UB/TIB Hanover and FIZ Karlsruhe, and the publishers are Physik Verlag, Kleinheim. Depending on the subject category, reports are available from UB/TIB, FIZ and from:

(1) University of Bonn Agricultural Sciences Library;
(2) Hanover Veterinary College Library;
(3) Central Library for Medicine in Cologne.

The arrangement of entries is as for *British reports translations and theses*, that is the modified COSATI subject category scheme.

Although the SIGLE database is uneven in national representation, the German-speaking countries are well served in respect of certain categories of grey literature by the *Gesamtverzeichnis der deutschsprachigen Schrifttums ausserhalb des Buchhandels (Bibliography of German-language publications outside the book trade)*, usually known as GVB, and published by K.G. Saur. The work covers the period 1966–80, and began publication in 1988. Eight volumes are planned, and the sources are the national bibliographies of Germany, Switzerland and Austria, and the

Forschungsberichte. Entries previously set out under various cataloguing rules are now arranged on a consistent pattern according to the *Regeln für die alphabetische Katalogisering* (RAK).

Postscript: UAP

No account of acquisition problems in relation to grey literature is complete without a reference to the Universal Availability of Publications (UAP) programme of the International Federation of Library Associations (IFLA). UAP is an objective and a programme developed by IFLA with the full support of Unesco. The objective is the widest possible availabilty of published material (defined as recorded information issued for public use) to intending users, wherever and whenever they need it, as an essential element in economic, scientific, technical, social, educational and personal development. To work towards this objective the programme aims to improve availability at all levels, from the local to the international, and at all stages, from the publication of new material to the retention of last copies, both by positive action and the removal of barriers. It has been seen as a major element in a wider concept of Universal Access to Information. UAP clearly has important implications for grey literature especially the linking of bibliographical control and availability at the national level, and the activities of the British Library Document Supply Centre within the framework of the programme have been reviewed by Vickers and Wood (1982). Developments are still unfolding and the opportunity to review the scene was taken at an international conference organized by IFLAs Office for International Lending in 1988 on the subject *Interlending and document supply*.

References

Aronson, E.J. (1986) *Report series codes dictionary.* 3rd edn. Detroit: Gale Research

Caponio, J.F. and Bracken, D.D. (1987) The role of the technical report in technological innovation. *5th International Conference of Scientific Editors*, Hamburg. PB 87–232500

Chillag, J. (1982) Non-conventional literature in agriculture – an overview. *IAALD Quarterly Bulletin*, **27**, (1) 2–7.

Committee on Scientific and Technical Information (COSATI) of the Federal Council for Science and Technology (1964) *Subject category and subcategory structure. AD612200*

Cook, J. (1985) *The price of freedom.* London: New English Library

Defense Documentation Center (1970) *Defense research and development of the 1960s: cumulated citations and indexes to defense-generated technical information – a user's guide.* Alexandria, V: DDC

Emrich, B.R. (1970) *Scientific and technical information explosion.* AD–717654

Godfrey, L.E. and Redman, H.F. (1973) Editors. *Dictionary of report series codes* 2nd edn. New York: Special Libraries Association

Hardick, M. (1987) *A guide to locating technical reports in US government publications collections.* ED 287506 Metrodocs Monograph One

Hay, A. (1988) Cloud on everybody's horizon. *Guardian*, 13 September

Klempner, I.M. (1968) *Diffusion of abstracting and indexing services for government – sponsored research.* Mehchen, N.J.: Scarecrow Press

McClure, C.R. Hernon, P and Purcell, G.R. (1986) *Linking the US National Technical Information Service with academic and public libraries.* Norwood, N.J: Ablex Publishing Corporation

Moore, L. et al. (1971) *Distinction is all – NTIS from a technical librarian's point-of-view.* ED–058913

Morita, I.T. (1988) *Current status of science and technology: grey literature in Japan.* PB88–227780/GAR

President's Science Advisory Committee (1963) *Science, government and information.* Washington: USGPO The Weinberg Report.

Rogers, C. (1987) HMSO, USGPO and the history of government publishing. *State Librarian, March, 5–8*

Vickers, S. and Wood, D.N. (1982) Improving the availability of grey literature. *Interlending Review,* **10**, (4) 125–130

CHAPTER THREE

Bibliographical control, cataloguing and indexing

Introduction

Bibliographical control – the identification and description of documents – may be exercised in a variety of ways and according to one of several systems, especially in the allocation to publications of uniquely identifying numbers. Grey literature has always been criticized for the complete absence or inconsistent application of any means of bibliographical control. Indeed it has often been remarked that one of the characteristics of grey literature is that any given document may assume several different guises, to the confusion of both user and supplier.

Bibliographical control

A system for the bibliographical control of books by means of book numbering has been in operation for many years and provides for the construction of an International Standard Book Number (ISBN) consisting of ten digits made up of the following parts: (1) group identifier (i.e. national, language, geographical or other convenient group); (2) publisher identifier; (3) title identifier; and (4) check digit.

The purpose of an ISBN is to identify one title or edition of that title if there is more than one, or volume of a multi-volume work, from one specific publisher, and it is unique to that title or edition or volume.

In the United Kingdom the administration of ISBNs is exercised by the Standard Book Numbering Agency Limited, which was set up by collaboration among J. Whitaker & Sons Limited (publishers of the trade journal the *Bookseller*), the British National Bibliography and the Publishers Association. The Agency's duties are to allocate identifiers to publishers and to see they create ISBNs satisfactorily; to allocate ISBNs in full to those titles which are published by organizations or persons not producing their own numbers; to register and record all ISBNs and to make them available for bibliographical and book listing services.

In recent years the EAN (European Article Number) has come into wide use, and can often be found printed on the back cover of books as a bar code with an eye-readable or machine-readable number in conjunction with it. An ISBN is incorporated into the EAN.

A system for serial publications has also been developed and involves the use of an International Standard Serial Number (ISSN) (Unesco, 1973) which is a seven-digit number plus a check digit written in the form XXXX–XXXX. The number is used to identify a serial title and is inseparably associated with this title. Any change in the title requires the allocation of a new ISSN.

Grey literature comprises a mixture of documents, some with ISBNs, some with ISSNs and many with neither. In the case of reports, the application of ISBNs has not been very common (examples are the various series of atomic energy reports announced by HMSO). In the United States most technical reports sent to USGPO depository libraries receive proper bibliographical control through their being indexed in the *Monthly catalog*. Many of the technical reports indexed in the *Monthly catalog* are also NTIS reports, but most NTIS reports are not depository, and so are not indexed in the *Monthly catalog*.

Over the years many systems of report numbering have been developed and refined, and in consequence the main means of bibliographical control in reports literature has been the report number, a simple concept which ought to result in report designations that are complete, consistent, concise and unique.

Regrettably because report numbering is a task assumed by those who handle reports as well as those who issue them in the first place, the situation surrounding report numbers has developed to the point where the Special Libraries Association has been obliged to use one word to describe it: *chaotic* (Godfrey & Redman, 1973). The remedy could lie in standardization and attempts have been made along these lines. However before examining what is involved in standardization, it is as well to see what happens when an issuing body decides to number a report. It sets out with the laudable intention of being thorough and so considers that a report ought to bear the following:

(1) Symbols for the name of the agency;
(2) Symbols for the subject matter of the report;
(3) Symbols for the form of the report;
(4) Symbols for the date;
(5) Symbols for the security classification;
(6) Symbols to show that additional data has been added by the recipient;
(7) Symbols which uniquely identify the report;
(8) Symbols which indicate the location within an organization.

The outcome of course is an excessively long number inconvenient to quote verbally or in writing and greatly susceptible to transcription errors such as the substitution of wrong letters, the transposition of numbers, and simple omissions or duplications. An example of how a report number grows is revealed by the following example, which in itself is not a report, but one of the many other categories of grey literature. The example is in fact a Dutch Standard NEN-3005 translated into English as NLL–M–20984 with a shelf location at Boston Spa of 5828.4F. In the United States, NASA decided to announce it in *Scientific and technical aerospace reports as* N72–28275 (NLL–M–20984–(5828.4F): NEN–3005).

The best agency to assign a report number is the issuing or originating agency. The next best are the contracting or assigning agencies – the trouble here is that a project may be sponsored by several agencies, each of which allocates its own identifier. Finally a report number may be assigned by a recipient. In national collections using widely recognized series, this procedure, far from being a problem, is a positive advantage in that often the recipient's number is all that need be quoted. Consistency demands that the various codes assigned at different stages in a report's progress be correlated, a reasonable enough requirement, except that in terms of file maintenance and cross-references the work involved can become excessively costly.

The use of an agreed uniform format for the creation of unique report numbers would enable issuing bodies to allocate numbers to their publications which would be compatible in arrangement with those assigned by others.

A standard is available for the numbering of technical reports (American National Standards Institute, 1983) wherein a technical report is defined as a document that gives the results of research or development investigations, or both, or other technical studies. The word technical is used to mean practice, method, procedure, theory, etc., in any science, art, business, trade, profession, sport, or the like. The Standard Technical Report Number (STRN) consists of two essential parts:

(1) A report code, to designate the issuing organization or corporate entity and in some cases a series or special series issued co-operatively by two or more organizations;
(2) A sequential group, characters which constitute that portion of the STRN assigned in sequence by each report issuing entity.

The maximum number of characters permitted in an STRN is twenty two (*Figure 3.1*), but an optional local suffix is also provided for; it is not part of the STRN, and may be of any length.

Nevertheless even if satisfactory provision is made for series of reports, translations and other documents issued in long runs, the problem of handing individual items from a myriad of originating sources still remains. Certainly the allocation of ISBNs offers one solution, but this requires a proper depository system, and the very nature of grey literature will preclude anything like a one hundred per cent operation of this principle.

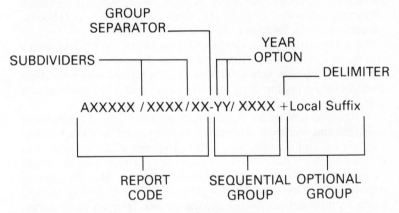

Figure 3.1 **Format of a Standard Technical Report Number** (STRN) (with acknowledgements to ANSI)

Cataloguing, indexing and retrieval

In one sense the cataloguing of grey literature does not require special treatment; it can be dealt with under the arrangements made for literature in general, of which it is but a small albeit growing part. However once again reports provide the exception, mainly because for many years they have been regarded as outside the mainstream of library material.

Where only small holdings of reports are concerned, such documents may conveniently be treated as part of a library's ordinary pamphlet collection. However, in cases where many reports are involved, practical problems arise owing to the wide range of information appended to each report. Standard library cataloguing rules may cease to be applicable and an organization must decide whether the time and effort involved are worthwhile and whether it is not more advantageous to adopt one of the sets of instructions issued by reports-providing agencies.

Some libraries find that their reports collections do not need cataloguing at all, owing to the thorough coverage provided by abstracting journals, and as a consequence simply make do with the provision of an elementary report serial number card or record for each item held. The main exception to this procedure will be the treatment of internal reports arising from the work of the staff in the library's parent organization.

Internally produced technical reports have long been a major resource of the company library. Such documents are written for internal circulation only and consist typically of engineering, production, sales development, process engineering and research department reports. Usually they include laboratory notebooks, progress reports, interim reports and final reports.

Thus a research organization will collect external reports, which can be treated in one of the ways outlined above, and it will also accumulate its own sub-species of grey literature which in effect constitutes a valuable corporate resource, and which will require some degree of cataloguing or other formal organization.

A standard scheme for the descriptive cataloguing of government scientific and technical reports has been evolved by COSATI and recently revised (Guidelines, 1986). The aim of the scheme is to provide rules for descriptive cataloguing appropriate to the needs of information centres, documentation centres and the reports departments of libraries; to provide users with a consistent form of citation and index entries among the various information systems; to enable government agencies to use each other's descriptive cataloguing entries with a minimum of editorial revision; and to provide a guide for other organizations. According to the scheme, the essential descriptive cataloguing elements are:

(1) Accession number;
(2) Corporate author;
(3) Title;
(4) Descriptive role – subtitle or progress report etc;
(5) Personal author;
(6) Date;

(7) Pagination;
(8) Contract number;
(9) Report number;
(10) Availability;
(11) Supplementary note;
(12) Security classification.

Each of the above elements is carefully defined and by far the largest amount of attention is devoted to the corporate author, described as the institutional or corporate body which has prepared and/or is contractually responsible for a report. Organizations most likely to be involved in issuing scientific and technical reports are identifed by COSATI as:

Academies
Arsenals
Associations
Business corporations
Centres
Colleges
Companies
Councils
Establishments
Firms
Foundations
Government agencies
Groups

Hospitals
Institutes
Institutions
Laboratories
Museums
Observatories
Proving grounds
Schools
Societies
Stations
Universities and their foreign (i.e. non-US equivalents).

The list is markedly similar to those drawn up by people attempting to define the scope of grey literature.

The corporate author is sometimes referred to as the source or *originating agency*. On the other hand the government or other agency which is financially responsible for the report and looks after its distribution is sometimes referred to as the *controlling* or *monitoring agency*. The scheme specifies that only two organizational elements may be chosen from those displayed on the title page and that they should be the largest followed by the smallest, e.g.:

Largest element: place name: smallest element
General Electric Co., Cincinnati, Ohio
Nuclear Materials and Propulsion Operation

Special rules apply for city and state names, departmental committees, abbreviations and change of name. Exceptions to the largest/smallest rule occur when one of the subordinate elements:

(1) Includes a proper name or a personal name: Bureau of Mines, Morgantown, W.Va, Appalachian Experiment Station;

(2) Is an independent name e.g. Atomics International, a division of North American Aviation Inc.;
(3) Is designated as the responsible organizational level by a report series number.

This emphasis on corporate authorship in reports literature is in marked contrast to traditional cataloguing procedures, where the personal author is supreme. This contrast is emphasized by the massive *Corporate author authority list* (Kane, 1987) prepared by NTIS. It is the printed version of the Corporate Author Authority Database created and maintained by NTIS, and contains 40 000 main entries for US Government sponsored research, development and engineering reports, as well as foreign technical reports and other analyses prepared by a whole range of agencies, their contractors and grantees.

Finally it should be mentioned that the original COSATI cataloguing scheme has been extensively developed and adapted by the Defense Technical Information Center (DTIC) (1988) in order to be able to specify the cataloguing information to be used in the data fields for the computer input of technical documents. The DTIC has also published a vocabulary, the DTIC *retrieval and indexing terminology* (DRIT) (1987) the aim of which is not only to be a manual for DTIC to index and retrieve information from its various data banks but also as to be an aid to assist DTICs users in their own information storage and retrieval operations.

It is possible to argue a case for the use of the Universal Decimal Classification (UDC) as a suitable scheme for the treatment of grey literature. It is after all one of the most detailed of the general classification schemes, particularly in the fields of science and technology; it is truly international, with versions in many languages; and its synthetic nature makes it amenable to use in computer systems. However UDC is not without its shortcomings and indeed its critics, especially in the vital area of revision, and whilst individual collections of grey literature may well be classified by it, the adoption of UDC by major announcement agencies does not appear to be imminent.

The *Anglo-American cataloguing rules* (AACR) can be used to deal with various kinds of reports under the section on works of corporate authorship, and do indeed recommend that a work that is by its nature necessarily the expression of the corporate thinking or activity of a corporate body be entered under that body. However it has to be remembered that the AACR is a set of rules for the construction of catalolgues of general collections and for that reason does not tend to provide for the cataloguing of specific categories of material (although it does cater for computer files).

The 1988 revision continues this policy and specialist material such as various categories of grey literature will have to be catalogued according to the general principles found in AACR.

A study comparing technical reports cataloguing records in the Library of Congress MARC format using AACR2 (i.e. 2nd edition) cataloguing rules and in the DTIC format using COSATI rules has been reported by Burress (1985), who notes that the automated systems and networks available for controlling monographs are not yet available for reports. Three solutions are possible:

(1) Treat reports as if they were monographs;
(2) Use the MARC communications format, inserting COSATI – derived information into hospitable fields;
(3) Develop special software.

The suggested COSATI format for catalogue cards is: *accession number, corporate author* (initial capitals), *title* (all capitals), descriptive note, *personal author* (initial capitals) date, pagination, *contract number, report number*, availability and supplementary note. The italicized elements are those which should be put to the left margin for emphasis and ease of marking for filing.

Of the descriptive elements recommended the following provide useful index entries in an abstract journal, or useful filing points in a manual catalogue: accession number; corporate author; title; personal author; contract number; report number. Examples of reports catalogued by this procedure are shown in *Figure 3.2*.

It is fairly easy to decide from the precise definition given in the COSATI scheme what an accession number actually is, but whether or not it is to be used for cataloguing purposes is not straightforward in cases where a document carries more than one accession number. For example many reports are identified by both AD and PB numbers.

A further aid to the cataloguing of reports is the common practice of including in each document a control sheet which requires basic information to be set out in a uniform manner. Standard instructions are issued for the completion of such control sheets, and the outcome is a thorough record which enables any cataloguing department to identify the key pieces of information needed for its records. An example of a control sheet as used by NASA is shown in *Figure 3.3*. At one time there was a practice of inserting in certain categories of reports and technical notes detachable abstract cards, for the convenience of librarians and others who had to maintain card files. Examples were to be found in the publications of the Royal Aircraft Establishment; the National Engineering Laboratory; and NASA.

Example 1

[1]PB-214 525/8
[2]Kentucky Univ., Lexington, Dept of Mechanical Engineering
[3]ENERGY TRANSFER IN FUR.   [4]Doctoral thesis.
[5]L. Berkeley Davis, Jr.   [6]Nov 72   [7]200 p.
[8]Grant NSF-GB-15579
[9]UKY-BU-101
[10]NTIS PC $3.00/MF $0.95

Example 2

[1]E73-10430
[2]Bendix Corp., Ann Arbor, Mich. Aerospace Systems Divn.
[3]ECOLOCIGAL EFFECTS OF STRIP MINING IN OHIO.
      [4]Bi-monthly progress report 1 Jan – 1 Mar 73.
[5]Phillip Chase.   [6]Mar 73.   [7]4 p.
[8]Contract NAS5-21762
[9]NASA-CR-131220
[10]NTIS PC $3.00/MF $0.95

    Examples of catalogue entries according to the COSATI
code: 1, accession number; 2, corporate author; 3, title;
4, descriptive note; 5, personal author; 6, date; 7, pagination;
8, grant or contract number; 9, report number; 10, availability.
Example 1 is a thesis supported by a grant; Example 2 is a
NASA Contractor Report and so qualifies for a NASA
Accession Number, N73-20391

Figure 3.2 **Examples of catalogue entries according to the COSATI
code**

1 Report No. NASA SP-7063	2 Government Accession No.	3 Recipient's Catalog No.
4 Title and Subtitle NASA Scientific and Technical Publications: A Catalog of Special Publications, Conference Publications, and Technical Papers, 1977–1986		5 Report Date September, 1987
		6 Performing Organization Code
7 Author(s)		8 Performing Organization Report No.
		10 Work Unit No.
9 Performing Organization Name and Address Office of Management Scientific and Technical Information Division Natinal Aeronautics and Space Administration Washington, DC 20546		11 Contract or Grant No.
		13 Type of Report and Period Covered
12 Sponsoring Agency Name and Add.ess National Aeronautics and Space Administration Washington, DC 20546		14 Sponsoring Agency Code
15 Supplementary Notes		

16

This catalog lists 2311 citations of all NASA Special Publications, NASA Reference Publications, NASA Conference Publications, and NASA Technical Papers that were entered into the NASA scientific and technical database during the decade 1977 through 1986. The entries are grouped by subject category. Indexes of subject terms, personal authors, and NASA report numbers are provided.

17 Key Words (Suggested by Authors(s)) Catalogs (Publications)	18 Distribution Statement Unclassified – Unlimited Subject Category 82		
19 Security Classif. (of this report) Unclassified	20 Security Classif. (of this page) Unclassified	21 No. of Pages 394	22 Price Free

NASA FORM 1626 AUG 87   Available from the National Technical Information Service,
                              Springfield, Virginia 22161 as PR 655B

Figure 3.3 **Example of a NASA worksheet** (with acknowledgements to NASA)

The question of indexing and retrieval systems for grey literature is one which concerns the user only insofar as it is necessary to understand the protocols required to address and exploit the various databases which cover the material in question. Major databases (for example the DOE *Energy database*: EDB; NASA STI; and NTIS *Bibliographic data file*) are considered elsewhere in this book. The main database which caters specifically for grey literature as a format rather than by subject is SIGLE (*System for information on grey literature in Europe*), available through BLAISE and INKA. However, given the patchy nature of the national inputs to SIGLE, the wideness of its subject coverage, and the general lack of appreciation of grey literature on the part of the educational, scientific, technical and academic communities, not to mention the library and information profession itself, SIGLE has a long way to go before it achieves the same sort of recognition and levels of usages as its longer established companions.

Filing

In determining filing practice the size of the collection of grey literature will be the deciding factor, and if only a few items are held, they can be filed by whatever method is convenient and consistent with the main collection of conventional items. If on the other hand the numbers are considerable, there are important advantages to be gained by segregation according to characteristics likely to be familiar to the reader. Thus runs of standards, sets of reports, collections of theses, files of transactions, and so on are best filed by whatever accession or other numbers have been allocated to them. There will be, of course, a requirement for an allowance of space between individual series to cater for additions to the collection, and extra complications arise if some of the items, mostly reports, are security classified. Here there is no option but to put them under lock and key in specially designated filing cabinets or even a properly designed strong room.

In establishing a filing system for grey literature the advantages of browsability should not be overlooked, even though in many cases the pamphlet like format means that items are unlikely to fit comfortably on open shelves. Where possible however grey literature items should be shelved alongside conventional publications. This applies for instance to certain long reports, especially those in hard or semi-stiff covers, which resemble ordinary books. A case in point is the NASA SP (*Special publication*) series where items like *Joining ceramics and graphite to other materials* (NASA–SP–5052, 84pp), *Advanced bearing technology* (NASA–SP–38, 511pp), and *Quest for performance: the evolution of nuclear aircraft* (NASA–SP–468, 548pp) are most useful to the reader if placed alongside similar texts in the ordinary literature.

Further practical guidance on the coding and filing of grey literature, in particular any official documents from whatever source, is contained in the work edited by Pemberton (1982), which gives detailed accounts of practice in libraries in Australia, Canada, Ireland, the United States and the United Kingdom.

Obsolescence

The period for which grey literature should be retained will vary with local conditions and requirements. Any policy will have to take into account the actual use made of the material, the availability of copies in archive collections elsewhere and the library's own role (if any) as an archive source itself. The use of microforms considerably eases the problems of storage and security.

Grey literature has not been around long enough as a distinct category to have given rise to any studies in its obsolescence, but certainly in the technical reports area a number of attempts have been made to determine how long publications can be usefully retained. In fact, high rates of obsolescence have been noted, particularly in the pioneer study by Wilson (1964) at the Atomic Energy Research Establishment; more recently Vickers and Wood (1982) were moved to say that 'it has been repeatedly mentioned that the demand for grey literature collected by the (British Library) Lending Division is very low'. Whilst a lack of awareness is undoubtedly one factor, and of course steps are being taken to correct it, obsolescence could be another.

Similar trends can be detected in the conventional literature, especially in the fall off in the use of periodical articles as the information they contain is absorbed into the general body of texts and reference works. A good example of this process is the classic manual *General and industrial management* by Henri Fayol (1841–1925), the principles of which were first published in a French mining journal. The work has since been through many editions to become a standard in its field, the latest revision appearing in 1988 (Fayol, 1988).

Conclusions

Apart from reports, the application of various forms of bibliographical control to grey literature has not until recently attracted a great deal of attention. Even here it has been noted that libraries will not find it exceptionally difficult to maintian bibliographic control over NTIS material (McClure and others, 1986), should it be purchased. The situation is of course due in a large measure to the lack of understanding of the nature and importance of the publications in question. For practical reasons bodies intimately concerned with reports in all their aspects have been obliged to seek their own solutions. To a considerable degree success has been achieved, but the results unfortunately stress the difference rather than the similarities between grey literature and conventional publications, and so to some extent perpetuate the feeling that the material is 'difficult'. Perhaps with the extra impetus provided by the Universal Availability of Publications (UAP) programme, the position may improve.

References

American National Standards Institute (1983) *Standard technical report number (STRN) – format and creation.* ANSI Z.39.23–1983.

Burress, E.P. (1985) Technical reports: a comparison study of cataloguing with AACR2 and COSATI. *Special Libraries*, Summer 187–192

Defense Technical Information Center (1987) Retrieval and indexing terminology, 3rd edn. AD–A 176 000

Defense Technical Information Center (1988) *Cataloging guidelines*. AD–A 192 200

Fayol, H. (1988) *General and industrial management*, revised by Irwin Gray. London: Pitman

Godfrey, L.E. and Redman, H.F. (1973) Editors. *Dictionary of report series codes*, 2nd edn. New York: Special Libraries Association

Guidelines for descriptive cataloging of reports: a revision of the COSATI standard for descriptive cataloging of government scientific and technical reports PB86–112349 (1986).

Kane, A.V. (1987) *Corporate author authority list* 2nd edn. Detroit: Gale Research

McClure, C.R., Hernon, P. and Purcell, G.R. (1986) *Linking the US National Technical Information Service with academic and public libraries*. Norwood, N.J: Ablex Publishing Corporation

Pemberton, J.E. (1982) *Bibliographic control of official publications*. Oxford: Pergamon (Volume II in the series 'Guides to Official Publications')

Unesco (1973) International serials data system *Unesco Bulletin*, **27** (2) 117–118

Vickers, S. and Wood, D.N. (1982) Improving the availability of grey literature. *Interlending Review,* **10,** (4) 125–130

Wilson, C.W.J. (1964) Obsolescence of report literature *Aslib Proceedings*, **16** (6) 200–201

CHAPTER FOUR

Report writing

Introduction

In a work concerned with reports as part of the grey literature it is not inappropriate to devote some attention to the practical and time-consuming task of writing a report, since those who are obliged to read other people's compositions are often called upon to prepare such documents themselves. Moreover any attempt to raise the standard of report writing ultimately confers a benefit on the report-reading community as a whole, for so often at the moment readers despair of particular reports because they do not convey enough information, or supply the wrong sort of information, largely as a result of the authors failing to think properly about the purpose of their work. Some guiding principles are in the interest of everyone.

English usage

Fundamental to all forms of writing is a good grasp of English usage, an attribute which applies just as much to reports as to any other type of composition. In the final analysis it can be argued that English usage is a very personal matter, almost but not quite dependent on an individual's taste. The fundamental requirement is that what is written conveys to all intended readers in a manner clearly understandable to them the information and opinions which the writer wishes to convey. In reports a literary style and the unusual use of words are best avoided. Some writers are able to express themselves correctly and lucidly in an intuitive fashion, whereas others welcome the support and comfort of a well-tried guide. Three such works which repay careful study are *Usage and abusage* (Partridge, 1970), a reference work arranged in dictionary fashion; the classic *A dictionary of modern English usage* (Fowler, 1983), revised edition; and the *Complete plain words* (Gowers, 1986), revised edition. The last-named, a widely acclaimed book which concentrates on what matters to the ordinary practitioner, and not on what interests only grammarians and scholars, is conventionally set out in chapters covering subjects such as the choice of words, the handling of words, correctness, punctuation, and recent trends in the development of the English language.

On the specific and specialized task of writing a scientific or technical report there are very many useful works, a large number of which are aimed at the undergraduate and college student. Some titles, however, are prepared on the premise that report writing is a duty which can fall to anyone at any stage of his or her career, and so acknowledge that instruction in the techniques involved needs to be directed at a wider audience than those people undergoing courses of full-time education. For example, *Writing technical reports* (Cooper, 1964) is a book based on experience gained in conducting report-writing classes, mainly for scientists and engineers, and an important point, which the author makes early on, is that often a person is sent on a report-writing course with the instruction to the course organizers, 'teach our delegate *how* to write', when what the delegate, really needs to be told is '*what* to write'. Equally down to earth is the approach adopted in the encyclopaedic two volume *Handbook of technical writing practices*, edited by Jordan (1971) in which the chapters on technical reports and on management reports deal exhaustively with content and structure.

Examples abound of style manuals and guides issued by government departments for the instruction of their own staff, and two recent instances are the *Technical report writer's style manual*, compiled by the US Army Tank Automotive Command (Boblenz, 1987) and the *Author–editor guide to technical publications preparation*, a handbook for the US Air Force Air Weather Service (AWS) (Horn, 1986). When one looks wider than just reports, a work which caters for other categories of grey literature is *How to write a research paper* (Berry, 1986); the book is aimed at students engaged on a fundamental course in higher education, and gives clear guidance on research papers, theses and dissertations.

At this juncture it is worth repeating just what a report is – an expository document which states in formal terms the results of, or progress made with an investigation or study, where appropriate draws conclusions and makes recommendations, and which initially is submitted to the person or body for whom the work has been done. Simply stated, the aim of any report is to influence the reader, to the extent of his or her taking necessary direct action, but more often merely to enable the reader to be better informed in making a decision which is affected by many factors and to which the report is just one contributing element. Consequently, before starting to write a report, the author will find it good practice to ask five fundamental questions:

(1) Who are the readers and can they be dealt with by just one report?
(2) What do they know already about the subject in question, and how much background information will they need?
(3) Why are they going to read the report: to find facts, ideas, recommendations, courses for action?
(4) How can they be induced to read the report in view of their possible apathy, indifference, ignorance or prejudice?
(5) When will they read the report and how much time will they be able to devote to it?

Planning the report

To deal with the issues raised by these questions, and so produce a successful report, requires very thorough planning. A *Specification for the presentation of research and development reports* (British Standards Institution, 1986), which incidentally stresses the point that no attempt is made to provide a manual of literary style, advocates the following steps as a start: (a) define the objective; (b) gather the information; (c) analyze the information; (d) draw conclusions; and (e) draft the summary.

Once the author has carried out this procedure, he must give due consideration to what he should write and how the work should be divided up. The title may not be the first consideration in the mind of the author, but since it is probably the first feature the reader will encounter, it is important that it indicates the subject matter clearly and concisely. Moreover in the case of a progress report the title should specify the exact period covered, together with the frequency of issue. A meaningful title is also very important in information retrieval systems using natural language keywords.

In a paper on the criteria for United States technical reports Hoshovsky (1965) provides a good example of a title which is a little too concise: it reads ELF INVESTIGATIONS, and is not, as might appear, an inquiry into the denizens of Middle-earth, but a document dealing with the subject of extremely low frequencies. Hoshovsky suggests that the essential elements of an informative title are:

(1) The object – what area is studied?
(2) The purpose – what are we looking for?
(3) The nature – report of an experiment, a state-of-the-art review, and so on.

Using this principle the title ELF INVESTIGATIONS is expanded into a much more helpful but hardly snappy 'World-wide magnetic field measurement of extremely low frequencies in the atmosphere'.

In similar manner, trouble should be taken in preparing informative abstracts – a task which unfortunately many report writers regard as a tiresome chore, to be avoided or delegated if at all possible. Briefly stated, the object of an abstract is to help in determining whether a report should be read in full, in part, or even at all. Its job is to describe the objective of the report, the work that was performed and the key results obtained, all within the compass of a few hundred words at most.

The body of the report proper should start with an introductory section, including a statement of the objectives of the activity reported and the reasons for starting it. Here too, the historical background should be explained. Depending on the exact nature of the report, the author should then go on to detail the theory where the report is primarily theoretical, or in the case of practical work describe the manner in which the experiments were carried out and the results obtained. In scientific and technical areas in particular it is helpful at this stage to indicate the experimental procedures used, with details of apparatus and techniques employed. In many cases it is necessary to expand the information on the experimental aspects, and this is best accomplished in an appendix or supplement (*Figure 4.1*), so that the main narrative flow is not interrupted. Tables and illustrations have the most impact when they appear at the most appropriate point in the body of the text; but where space considerations or reprographic constraints make this impossible, they too should be collected to form an appendix.

Knowing just how deeply to go into a subject calls for a nice sense of judgement, and as a good general principle the amount of detail given should be just sufficient to enable an adequately skilled worker in the field to retrace the steps of the investigation or study without too much difficulty, although clearly this will depend very much on the author's conception of his readers. The safest approach is to write to the level of the least knowledgeable readers, even though such a step involves some sacrifices in

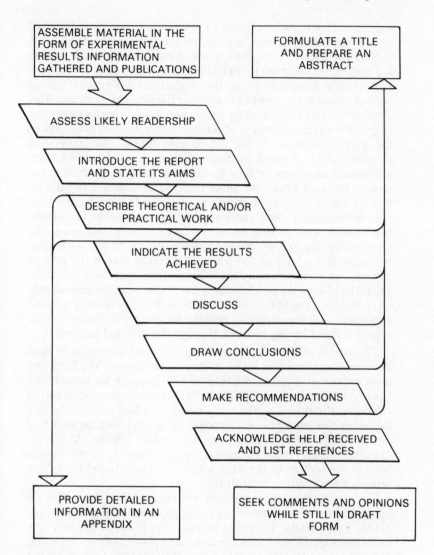

```
ASSEMBLE MATERIAL IN THE          FORMULATE A TITLE
FORM OF EXPERIMENTAL              AND PREPARE AN
RESULTS INFORMATION              ABSTRACT
GATHERED AND PUBLICATIONS

ASSESS LIKELY READERSHIP

INTRODUCE THE REPORT
AND STATE ITS AIMS

DESCRIBE THEORETICAL AND/OR
PRACTICAL WORK

INDICATE THE RESULTS
ACHIEVED

DISCUSS

DRAW CONCLUSIONS

MAKE RECOMMENDATIONS

ACKNOWLEDGE HELP RECEIVED
AND LIST REFERENCES

PROVIDE DETAILED                 SEEK COMMENTS AND OPINIONS
INFORMATION IN AN                WHILE STILL IN DRAFT
APPENDIX                         FORM
```

Figure 4.1 **Recommended sequence of events in report writing**

conciseness. Whatever the level at which the narrative is pitched, the actual impact is greatly enhanced if the material is divided into a logical sequence of sections and sub-sections. When the author has stated what has been accomplished, he must then discuss it in order to provide an interpretation of the outcome (or a commentary on it) and to supply the reasoning on which the conclusions are founded. This is also the place to illuminate the work in terms of new or extended theories or principles in the field covered.

Next the author must draw conclusions, a commitment writers are sometimes reluctant to undertake. The writer must give a clear and orderly presentation of the deductions made after a full consideration of the results obtained or the findings gathered. This should then be rounded off by recommendations, preferably in the shape of concise statements of further action deemed necessary as the logical outcome of the conclusions drawn. Naturally such recommendations ought to derive directly from the conclusions and should moreover be fully justified by the work covered in the report. In cases where the proper course of action is to take no further action the author should not be afraid to say so.

It is customary, and indeed perfectly reasonable to present conclusions and recommendations immediately following the sections devoted to experimental procedures, results and discussion, but there is a school of thought which argues that in the case of reports intended especially for management, there is much to be gained by placing the conclusions and recommendations immediately after the introduction. Thus is created the so-called 'executive summary' and a manager is enabled to grasp the essence of the report without having to work through the detailed sections.

The report may be completed with acknowledgements of help received and also with a list of references or sources. The latter are most important if the reader is to be able to check for himself any books, reports, or papers cited, and accordingly the form of reference should be a bibliographically standard one, preferably following the precepts of a recognized system such as that laid down by the British Standards Institution (1976). A further important point is that references ought to appear in the correct form at the bottom of the page where first cited, to aid in reading reports transferred to microform.

Even with the best-organized material and a clear plan of approach, the task of actually putting thoughts on paper does not always come easily. It cannot be over emphasized that very few individuals are blessed with the combined powers of concentration and expression to such an extent that they are able to dash off the final copy in the first draft. It is as well, therefore, to be reconciled to the fact that writing can be a painful business, consisting largely of a sequence of draft and revise. There is without doubt a great deal to be said for getting into some sort of written form (or on to a tape in a pocket dictating machine) as quickly as possible the substance of the flow of ideas as they emerge from an intensive and immediate consideration of the material to hand; but, once this stage has been achieved, it is often very productive to pause and reconsider even to the extent of leaving the task of writing altogether for a period of time, in order to be able to return with a fresh and critical mind.

Polishing the draft

When an author is satisfied with the draft, the next step ideally ought to be to give it as much polish as possible, for the final version will be judged not on the effort which went into the preparation but on the impression it leaves on the reader. Desirable as polish is however, it is not always attainable, frequently because of outside pressures such as a fixed publication date or the need for the author to give attention to other equally urgent matters. When there is time for polishing it should be devoted to achieving the greatest possible clarity of expression by attention to such features as sentence and paragraph construction.

Of great value too at this stage of a report's progress is a second opinion, perhaps from a colleague or more probably from someone in an official capacity such as an editor concerned with standardizing the level of content and presentation throughout a whole series of reports or other documents.

It is at this point that a clear indication emerges as to whether the writer has succeeded in the prime task of imparting information in such a manner as to influence the reader. The great value of a second opinion is that it draws attention to some of the besetting sins of report writers, one of which is an over-fondness of the use of initials and acronyms without ever pausing to explain them.

As a golden rule, all abbreviations, signs, symbols and units should comply with generally accepted and universally recognized systems. At the first instance of their use in the text, abbreviations and acronyms should be accompanied by an explanation of their meanings. Another shortcoming in report writers is a tendency to confuse issues by using several different terms to describe the same concept – in other words, inconsistency of nomenclature. Once again standards can help, and in particular the many glossaries issued by the British Standards Institution are quite invaluable. A complete list is given in the *BSI Yearbook*. Finally some report writers have a leaning towards too much brevity by taking for granted that readers are not in need of an explanation of specific points when in reality they are.

Another important aspect which should be noted is that although the actual printing and distribution of a report, indicating such matters as report numbering and the allocation of any security or confidentiality classification are not likely to be the prime responsibility of the author, it is nevertheless an advantage to be familiar with an organization's standard procedures governing the preparation and distribution of publications in general. Often such procedures are codified in internally produced handbooks and manuals. Alternatively it is good practice to follow a national standard such as the British Standard mentioned earlier. In the United States report writers can follow the American National Standard Institute (1974) *Guidelines for the format and production of scientific and technical reports*, which prescribes the order and specifications of the elements of a report, including a standard report documentation page.

Whichever standard is used, there is a book which gives excellent guidance on many questions which arise in preparing printed, and for that matter typewritten or word-processor produced documents; it is the *Oxford dictionary for writers and editors* (Oxford Dictionary Department, 1981), the successor to eleven editions of the well-known Collins *Authors' and printers' dictionary*. Spelling, punctuation, italicization, capitalization, abbreviation, foreign words and phrases, and printing technicalities are all dealt with in detail. The advent of spelling checks in personal computers has made the life of the writer a lot easier, but pitfalls still abound – does one mean discreet or discrete, for example?

It is also particularly important to check on specifications laid down for the reproduction of photographs, line-drawings, tables, graphs and other illustrative material. Whilst it may be true that every picture tells a story, unless that picture can be satisfactorily reproduced, the author will be forced to devote more space than had been anticipated simply to provide an adequate description.

Another aspect to heed is the way an organization handles the presentation of data available only in the form of computer output. The temptation is to use computer output in the form in which it emerges from the printer, albeit consigning it to an appendix, but often a far better result is obtained by having the data, or at least key sections of it, retyped or reformatted in a conventional style.

Conclusions

Sad to say many of the reports and other documents discussed elsewhere in this book come nowhere near meeting the basic requirements outlined above, but that in itself is no reason to deter writers from adopting a methodical approach and adhering to good standard practice.

References

American National Standards Institute (1974) *Guidelines for format and production of scientific and technical reports* ANSI Z39.18 1974. (Note – The former Z39 committee of ANSI has become independent under its new name, National Information Standards Organization Z39 NISO)

Berry, R. (1986) *How to write a research paper*, 2nd rev edn. Oxford: Pergamon

Boblenz, J.N. and Calhorn A.A. (1987) *Technical report writer's style manual.* Report AD–A177791

British Standards Institution (1976) *Recommendations: bibliographical references.* BS 1629: 1976

British Standards Institution (1986) *Specification for the presentation of research and development reports.* BS 4811: 1972 (1986)

Cooper, B.M. (1964) *Writing technical reports.* Penguin (1964 and many reprints)

Fowler, H.W. (1983) *A dictionary of modern English usage*, 2nd rev edn. by Sir Ernest Gowers. Oxford: OUP

Gowers, Sir E. (1986) *The complete plain words*, revised by Sidney Greenbaum and Jane Whitcut. London: HMSO

Horn, G.M. *Author–editor guide to technical publications preparation. Report AD–A192 850 (1986).*

Hoshovsky, A.G. (1965) *Suggested criteria for titles abstracts and index terms in Department of Defense Technical Reports.* Report AD–622 944

Jordan, S. (1971) (Ed.) *Handbook of technical writing practices.* New York: Wiley-Interscience

Oxford Dictionary Department (1981) *Oxford dictionary for writers and editors.* Oxford: Clarendon Press

Partridge, E. (1970) *Usage and Abusage: a guide to good English.* Penguin (1970 and reprints)

CHAPTER FIVE

Theses, translations and meetings papers

Introduction

The term grey literature by no means meets with universal acceptance, and in the area of theses, translations and meetings papers, three categories of publication with long and honourable pedigrees, the description sits particularly uneasily. The justification for the inclusion of theses, translations and meetings papers in the first edition of this work, devoted then entirely to reports literature, rested on the following factors:

(1) Theses, translations and meetings papers are quite frequently announced in journals devoted mainly to reports, and so come to be identified with but not regarded as types of reports literature;
(2) Translations, theses and meetings papers have some of the format characteristics of reports proper;
(3) Theses, translations and meetings papers are commonly assigned identifying numbers, either by issuing establishments or collecting agencies, and such numbers bear strong resemblances to those applied to reports;
(4) Theses, translations and meetings papers are regarded, not without some justification, as difficult materials to identify and obtain, problems frequently mentioned in connection with reports.

These factors still apply and the justification for continuing to treat theses, translations and meetings papers as a separate sub-group is undiminished.

Theses

A definition of a theses formulated by the British Standards Institution (1986) is a 'statement of investigation or research presenting the author's findings, and any conclusions reached, submitted by the author in support of his canditature for a higher degree, professional qualification or other award'. It is immediately apparent from this definition that a thesis (in the United States the preferred term is 'dissertation') has several points in common with a report. Both present details of investigations and/or research; both offer findings and conclusions; both are submitted to an overseeing body (the commissioning agency or the university); and both are regarded as unpublished documents, except, as noted below, in the case of some European universities.

British theses

In the United Kingdom universities have generally adopted a very conservative attitude to the provision of copies of theses, in marked contrast to the arrangements prevailing in the United States and Europe. There is no central supply channel for British doctoral theses, although the British Library Document Supply Centre has succeeded in creating a comprehensive central collection going back to 1970. Additions are publicized in *British reports theses and translations* and currently BLDSC receives theses from virtually all UK universities and the Council for National Academic Awards, who supply their doctoral material for microfilming. In 1988 BLDSC held 76 000 doctoral theses, and its annual intake was a further 6000. The Centre also has 416 000 US dissertations, again with an annual intake of 6000.

An important reference work for the verification of details about British theses, whether held at Boston Spa or not, is the *Index to theses with abstracts accepted for higher degrees by the universities of Great Britain and Ireland and the Council for National Academic Awards*, compiled and published by Aslib on a regular basis since 1953. From the *Index . . .*, which spans the period from 1950 onwards, it is possible to ascertain what theses have been submitted and then apply directly to the universities in question. From volume 35 onwards, the *Index . . .* was published in an expanded and improved version to include full texts of abstracts and a greatly enhanced subject index. Volume 36 appeared in 1988, and covers in the main theses submitted in 1985 and 1986. Around 9000 theses a year are indexed. For British theses accepted before 1950, checks can be made in the *Retrospective index to theses in Great Britain and Ireland 1716–1950*, a comprehensive five-volume work published in 1975. A further development in cumulative indexes is the *Brits index*, an index to British theses collections (1971–1987) held at the British Library Document Supply Centre and London University. The *Index* was published in 1988 in three volumes as a co-operative venture between the British Library and Information Publications International (IPI).

Availability will vary from university to university, but many have agreed to follow the procedure suggested by the Standing Conference of National and University Libraries (SCONUL), which states: (a) that at least one copy of every thesis accepted for higher degrees should be deposited in the university library; (b) that, subject to the authors' consent, all theses should be available for inter-library loan; (c) that, subject to the authors' consent, all theses should be available for photocopying; and (d) that authors of theses should be asked at the time of deposit to give their consent for (b) and (c) in writing, and that this consent should be inserted in the deposit copies of theses.

Finally it should be noted that many universities require the reader of a borrowed copy of an unpublished thesis to sign a declaration that no information derived from a study of the text will be published or used without the consent in writing of the author.

United States theses

In the United States and Canada, and indeed throughout the world, nearly five hundred co-operating institutions ranging from the University of Adelaide to York University, Canada, submit either the full texts or abstracts of doctoral dissertations to the Ann Arbor based University Microfilms International (UMI), who then prepare a monthly compilation called *Dissertation abstracts international* (DAI). This publication appears in three sections, namely:

Section A – the humanities and social sciences;
Section B – the sciences and engineering;
Section C – European abstracts (published quarterly).

Each author-prepared abstract, up to 350 words in length, describes in detail the original research project upon which the dissertation is based. Most of the approximately 30 000 dissertations published by UMI each year are abstracted in *DAI* and may be purchased in microform or as paper copies. As with many other databases the importance of CD-ROM (Compact Disc-Read Only

Memory) products is becoming apparent, and the *Dissertation abstracts* database is also available on two archival discs covering the period 1861 – June 1984 and July 1984 – 1987.

Other printed reference books prepared by UMI include the 37 volume *Comprehensive dissertation index* (CDI) covering the period 1861–1972, which lists more than 417 000 dissertations accepted in North America under keyword headings with a separate author index. The *CDI Ten year cumulation 1973–1982* cites nearly 351 000 dissertations in a 38 volume single source reference to *current* dissertation information.

Masters abstracts international, published quarterly, provides 150 word author-prepared abstracts of masters theses, whilst *Research abstracts*, also published quarterly, includes 300 word abstracts of post-doctoral and non-degree published research in special areas such as psychology and education. Finally *American doctoral dissertations*, issued annually on an academic year basis is arranged by subject categories, institutions and authors, and compiled for the Association of Research Libraries.

UMIs dissertation information service is in fact a programme of publishing, bibliographic and copying activities, the foundation of which is a comprehensive database accessible by a search system called Datrix Direct, and complemented by *Dissertation abstracts online*, which reproduces the full text of abstracts published since July 1980.

An analysis of the doctoral dissertation's role as an information source by Boyer (1972), appropriately enough in the form of a doctoral thesis, noted that while doctoral dissertations must embody the results of extended research, be an original contribution to knowledge and include material worthy of publication, they are often overlooked in the compilation of bibliographies and reading lists.

Theses in other countries

In order to trace theses in other countries it is necessary to consult the national lists published in the countries concerned. French theses have been recorded since 1884 in the *Catalogue des thèses de doctorat soutenues devant les Universités Françaises* (the exact title varies), and also in *supplement D* of *Bibliographie de la France*. German theses may be traced through the *Jahresverzeichnis der deutschen Hochschulschriften*, which changed its title with volume 86, 1970 to *Jahresverzeichnis der Hochschulschriften der DDR, der BRD und West Berlin*. For German language academic publications as a whole it is necessary to consult the *Gesamtverzeichnis deutschsprachiger Hochschulschriften* (GVH) 1960–1980, and its predecessor, the *Gesamtverzeichnis des deutschschprachigen Schrifttums 1911–1965*. GVH was published by K. G. Saur in twenty-four volumes between 1984 and 1987, and contains details of dissertations, post doctoral theses and university publications from West Germany, East Germany, Switzerland and Austria. It also includes information on German language academic papers published abroad, for example Australia, Canada, the USA and several other countries. A sixteen volume index is due to appear in 1989.

In addition to such comprehensive lists, it is possible to examine the lists issued individually by several continental universities – for example *Disertationen: Rheinisch–Westfalisch Technische Hochschule, Aachen*, and *Theses:Université de Geneve Faculté des Sciences*. Copies of the items listed in many such individual compilations are in the first resort offered for public sale by publishers appointed by the universities, and so take the material right outside the bounds of grey literature.

Translations

Translations have always been regarded by some as part of the reports literature, and more recently by practically everyone as part of the grey literature. In fact, of the thousands of translations made each year, relatively few are of reports; mostly the originals in question are ordinary books, papers, standards and patents. A closer look however reveals that apart from certain entire volumes and some cover-to-cover translations of journals, most translations are of individual items such as papers from journals, articles in newspapers or sections from books, and in the translated versions many of these items have several of the attributes of report noted earlier: pamphlet format, identification numbers, issuing agencies and also inconsistency of bibliographic control. Moreover several of the large agencies which have assumed responsibility for collecting, co-ordinating, indexing and publicizing translations are also the same agencies responsible for dealing with reports, or are organizations which are run on similar lines to reports agencies and have adopted their methods. An illustration of this is the fact that the National Translations Center (see below) has utilized the COSATI subject category list to arrange its translation register.

Translations have thus always been an important part of the reports literature and are a major constituent of the grey literature. The reasons why translations, especially technical translations, are in such steady demand are not hard to seek, and may best be summed up by reference to some of the conclusions reached after a classic survey conducted by the National Lending Library and reported by Wood (1967): (a) something like 50 per cent of the world's scientific and technical literature is published in languages other than English; (b) information contained in these foreign languages is vital for the work of English speaking scientists around the world. The study goes on to recommend that published guides and unpublished indexes to translations should be given more publicity, and some organizations have taken up the challenge, as will be seen below.

Before the arrangements for the translation of individual items (papers, parts of books, standards, patents, etc.) are examined, attention must be turned to an area where bibliographical control is especially good, namely translated versions of entire individual books and cover-to-cover translations of journals.

Translated books are listed annually in *Index translationum*, published by Unesco. The index enables the reader to follow, from year to year, the flow of translations from one country or central region to another and assists in tracing works by a specific author as they appear in translation. Volume 35, which was published in 1988, contains information on translations issued in the year 1982.

A point worth noting with regard to the various journals which announce new translations is that they are tertiary services in as much as the scientist or research worker who is seeking information will have done his searching in the conventional primary journals and secondary sources such as abstracting publications. It is unlikely that he will be interested in seeing what translations are of use to him by browsing through the appropriate subject categories. It is far more likely a worker will get to know of the foreign language papers and reports of interest to him through contacts with colleagues, especially those overseas, and his main concern will be to confirm that they are available in translation. For that reason translation journals are the special province of the librarian and documentation expert, who have the task of making sure the contents of such publications are made known to potential beneficiaries.

The pros and cons of cover-to-cover translations have been rehearsed many times – on the one hand they are wasteful because they include many articles of doubtful value; on the other hand they are a means of avoiding omissions and certainly simplify checking when searching an extensive file. Evidence of the system's durability is the fact that over 1100 cover-to-cover and selective translation journals have been produced, mostly from Russian originals, but also from other Slavonic languages and from Japanese, Chinese and German (see *Figure 5.1*). Full details may be found in *Journals in translation*, a guide to periodicals

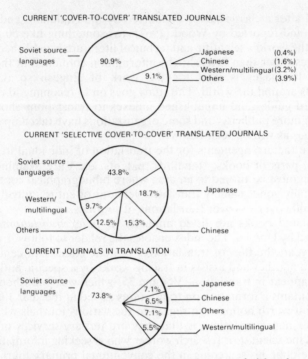

CURRENT 'COVER-TO-COVER' TRANSLATED JOURNALS

Soviet source languages — 90.9%
9.1%
Japanese   (0.4%)
Chinese   (1.6%)
Western/multilingual (3.2%)
Others   (3.9%)

CURRENT 'SELECTIVE COVER-TO-COVER' TRANSLATED JOURNALS

Soviet source languages — 43.8%
18.7%
Western/multilingual 9.7%
Others 12.5% 15.3%
Japanese
Chinese

CURRENT JOURNALS IN TRANSLATION

Soviet source languages — 73.8%
7.1% Japanese
6.5% Chinese
7.1% Others
5.5% Western/multilingual

Figure 5.1 **Sources of translations by language** (with acknowledgements to ITC)

translated cover-to-cover, abstracted publications and periodicals containing selected articles; the fourth edition was published jointly by the International Translations Centre and the British Library Document Supply Centre in 1988. On a more selective basis, a list called *Japanese journals in English* was published in 1985 by the then British Library Lending Division and the British Library Science Reference Library: it covers the location of scientific, technical and commercial periodicals.

An important factor determining the usefulness of cover-to-cover translations is the time-lag between the publication of the original and the appearance of the translated version; in some cases the gap can be so great as to warrant commissioning the translation of particularly important papers in advance of the cover-to-cover text.

Many thousands of translations of individual pieces of literature are published each year by government departments and agencies, by industrial firms, by universities, and by commercial translators. In the United States the major announcement publication and retrieval tool was for several years *Technical translations* (1959–1967), issued by the Clearinghouse for Federal Scientific and Technical Information (now NTIS). Then a change in policy resulted in translations originating in the US Government sector, such as the TT series sponsored by the National Science Foundation's Special Foreign Currency Science Information Program, and the JPRS series, discussed below, being announced in the publication *US Government research and development report* (now GRA & I); while the National Translations Center, a co-operative non-profit enterprise which was founded in 1963 after having existed for a number of years as a volunteer project of the Science and Technology Division of the Special Libraries Association, became a depository and information service for translations originating in the US non-government sector and began publishing the *Translations register-index*.

The *register* section of this announces new accessions in the Center. The *index* section includes these accessions as well as items listed by NTIS in GRA & I, and items available from commercial translation agencies and many other sources. The *index* sections of *Translations register-index* accumulate quarterly for all entries to date included in a volume. For NTC accessions the *register* section lists translations in subject categories according to the COSATI terminology and the final units of the identifying reference number correspond to the COSATI classification. Translations available from NTC are also available from British Library Document Supply Centre on loan or in photocopy. Reciprocally, translations by BLDSC are available from NTC to applicants in the western hemisphere.

In 1969 the NTC issued an important guide, *Consolidated index of translations in English* (CITE) which cumulates translations added to the pool prior to 1966 and consolidates translations announced in *Technical translations; bibliography of translations of Russian scientific and technical literature (1953–1956)*; the SLA *Author list of translations (1953)* and *supplement (1954)*; *Translations monthly (1955–1958)*; and *Bibliography of scientific and industrial research (1946–1953)*. Altogether *CITE* contains details of 142 000 translations. Its companion *CITE II*, covering the year 1967–1984, appeared in 1987. In all the National Translations Center contains information on over one million translations, nearly a half of which comprise the NTC collection.

However, for users who do not have access to such major bibliographic tools a useful guide has been published to translations of scientific and technical literature by the Naval Ocean Systems Center (Wright, 1987); this describes major providers of existing translations and suggests procedures for having a publication translated by a commercial firm or a government agency.

A particularly intensive area of translating activity in the United States, which has been in progress for many years, is that conducted by the Joint Publications Research Service (JPRS), which produces many hundreds of pages of scientific and technical translations each year, a large percentage of which were originally published in the Soviet Union. Translations are also made of political and economic material from China, North Korea, Eastern Europe, Africa, Latin America and the Middle East.

JPRS is a part of the US Department of Commerce and was established to provide translations support to the US Government. Its publications are composed of items selected from foreign language sources by various government agencies and departments, and translated and printed by JPRS. Translations are issued on an ad hoc basis, for which reference aids are available (e.g. JPRS 58548) which list ad hoc publications issued in specific subject categories during a given year. They also appear as regular series in the form of English translations of publications from various geographical areas – for example *Science and technology: Europe and Latin America* (JPRS–ELS–88–001 etc.). JPRS translations are notified in *Government reports announcements and index*. Further access to JPRS publications is discussed in an appendix to ITCs *Journals in translation*, and the aims and objects of the translations programme have been assessed in a paper by Morton (1983). The main announcement publication is *Transdex index*, a monthly index in paper format with an annual cumulation on microfiche. It is issued in upper case computer print-out and comprises:

(1) Series and ad hoc title section;
(2) Bibliographic section;
(3) Keyword section;
(4) Personal name section.

Great Britain does not have a translations centre so named but until 1984 Aslib maintained a card index of existing and in-progress translations. Known officially as the *Commonwealth index of unpublished scientific and technical translations*, the service offered a purely location facility with records of close on half a million translations from all languages into English.

The success rate in establishing locations was not particularly high, about ten per cent. A valuable adjunct to the *Index* was an Aslib service providing the names of translators expert in given languages and also possessing a knowledge of specific subject areas. In 1984 the *Index* was merged with the database operated by the International Translations Centre in Delft and is now available to users through the Centre.

The British Library began collecting translations and details of translations in the late 1950s and the records of holdings now total over 500 000 including translations prepared by the National Technical Information Service, the National Translations Center and JPRS. The British Library Document Supply Centre actively encourages British government departments, firms, institutions, research associations, universities and other bodies to contribute translations to its collection. Several thousand items per year are received from British sources. BLDSC accepts queries about the availability of translations and if it is unable to locate a source, it will check with other agencies.

At one time the library published the *LLU (Lending Library Unit) bulletin*, later to become *NLL translations bulletin*, but subsequently as a result of arrangements between the National Translations Center and the British Library, an exchange of translations and information about translations was established between the two centres. Translations acquired by Boston Spa (BLDSC) from British sources are listed in the index section of NTCs *Translations index register*. Current announcements of British translations are also reported in *British reports theses and translations* and documented in BLDSCs *Translation index* (TI), a major cumulated card index.

Another example of a national initiative is the *Canadian index of scientific translations* maintained by the Canada Institute for Scientific and Technical Information (see Chapter 2), which gives the locations of over 500 000 translations of foreign language documents.

For many years the British Library (BL) operated its Russian Translating Programme whereby translations of Russian papers and books (generally not more than two years old unless considered of special significance) were translated into English free of charge provided the organization submitting the request undertook to edit the manuscript for technical correctness. Subsequently the translations appeared in the RTS series and were announced publicly, at which stage interested parties could purchase copies at very low cost.

In 1973 BL initiated a modification to its translation service by asking enquirers whether an English language summary of the paper in question had already been seen. The object was to establish whether translation requests could be satisfied in the first instance by provision of an English language summary or translated captions and conclusions, plus a photocopy of the original item. The experiment showed that English summaries plus short notes were acceptable in many cases, and that perhaps a great deal of unnecessary translating activity is taking place.

Currently the British Library Document Supply Centre sponsors ten cover-to-cover titles produced by research associations and academic institutions.

Many smaller specialized translation indexes are also maintained in the United Kingdom – for example by the UKAEA, the Lead Development Association and the Central Electricity Generating Board. The Institute of Metals has a translations collection of over 25 000 items, including the well-known BISI series, and in 1987 added a further 1100 translations to the total.

In Western Europe, (apart from the British Library Document Supply Centre) the largest translations enquiry service is run by the International Translations Centre (ITC) (formerly the European Translations Centre). ITC is a not-for-profit international network inaugurated in 1961 under the auspices of the Organization for Economic Co-operation and Development (OECD). Its aim is to prevent the duplication of translating effort by collecting processing and disseminating information on existing translations in all fields of science and technology. It provides access to translations through the already noted *Journals in translation*, and through *World translations index*, a database and printed publication contained 280 000 references jointly produced by ITC and CNRS-INIST (Centre National de la Recherche Scientifique – Institute de L'Information Scientifique et Technique), Paris, in co-operation with US National Technical Information Service. ITC began by concentrating on translations from non-western languages into western languages, but has gradually been increasing its source-language coverage and now adds to its files details of translations from one European language to another.

Printed cumulations of ITCs records are available as *World transindex* (1977–1985) and *World index of scientific translations* (1967–1971 and 1972–1976).

On the question of the availability of translations, some organizations specifically forbid reproduction by photocopying and insist on outright purchase. This is because translating is an expensive and labour-intensive activity, which is likely to remain so until machine produced translations become available.

Finally, despite the considerable efforts made around the world to keep track of and make accessible the translations that have been made, enquiries to national centres are more likely to result in disappointment than success. The only recourse then is to commission a translation, either using in-house staff if the organization is a large one, or seeking advice on suitably qualified translators from bodies such as Aslib, which still maintains the register mentioned earlier.

A useful guide to the presentation of translations is to be found in BS 4755 (British Standards Institution, 1983), and of course when completed a copy of the translation should be added to the general pool of grey literature by notifying the details to the British Library Document Supply Centre.

Meetings papers

'Meetings papers' or 'preprints' are terms usually applied to papers made available in advance of or at meetings and conferences, where they are presented by their authors. The practice is especially common in the United States and many large technical and engineering societies issue preprints in advance of their meetings, with each paper bearing an identification in the form of a serial code. After the meetings have taken place, the papers are critically reviewed and a certain proportion selected for inclusion in the society's permanent records. The remainder are simply listed, and although they are not intended to be a part of the permanent literature, they are nevertheless in the public domain and so quoted and requested. An irritating complication is that not all papers announced before an event and assigned identification numbers are actually issued; some may be presented orally and then withdrawn, some never presented at all.

Preprints and meetings papers undoubtedly qualify for inclusion in the grey literature, but ought not to be confused with reports, since quite clearly they are (or could be) advance copies of journal or transactions papers. However because they carry numbers which resemble in some respects report numbers and because they too have a pamphlet format, many users of the reports literature tend to regard them as reports as well. Abstracting and indexing services are well aware of the distinction, and major services such as *Engineering Index* treat them as part of the conventional serial literature.

Not all the societies which issue preprints can be covered in this section; instead the characteristics of some of the series most frequently met with will be examined.

American technical societies

The American Society of Mechanical Engineers (ASME) issues a widely used series of preprints, which are announced in the Society's journal *Mechanical engineering* and in its programmes for specific meetings. Preprint numbers indicate meetings by date and code, e.g. paper no. 88–WA/APM–22 means paper no. 22

presented at the Winter Annual Meeting 1988, and contributed by the ASME Applied Mechanics Division. A selection of papers is published in the *Quarterly transactions of the ASME* and *Annual indexes to ASME papers*, included in the bound *Transactions of the ASME*, list all papers, even those not selected for the *Transactions*.

The Society of Automotive Engineers (SAE), now subtitled the Engineering Society for Advancing Mobility: Land Sea Air and Space, also has a very active publishing programme, the thinking behind which has been explained by Staiger (1973). SAE meetings papers are announced in one of the Society's two news journals *Automotive Engineering* and *Aerospace Engineering*, and in the programmes for specific meetings. Paper identification takes the form SAE 880574 – *A lean-burn catalytic engine*, where the first two digits represent the year of presentation. Just under half, after careful review by SAE committees are selected as having the greatest long term reference value and published in the normal way in *SAE transactions*. The rest stay firmly in the grey literature category, but complete sets of published and unpublished technical papers are made available on microfiche, and individual titles of either category can be ordered as required. A cumulative subject/ author/chronological index dating from 1965 is revised on a regular basis; the 9th edition covering the period 1965–1987 and giving details of over 22 000 technical papers was issued in 1988.

Other societies which publish meetings papers on a regular basis include the Society of Manufacturing Engineers (SME) with approximately 800 new papers each year; the American Institute of Aeronautics and Astronautics (AIAA), the papers of which are announced in the Institute's *Journal*; the American Society of Lubrication Engineers, which uses its journal *Lubrication engineering* to announce its preprints; and the vast output of the American Society for Testing and Materials (ASTM), information about whose papers can be found in the monthly *Standardization news*.

Before and after

Information about forthcoming meetings and conferences is available from a very wide range of sources, and listings appear in daily newspapers (for example the *Financial Times*) and in many weekly and monthly journals. In the United Kingdom a major co-ordinating service is *Forthcoming scientific and technical conferences*, a long-established compilation produced by Aslib each year with quarterly updates. Events are listed in chronological order, looking ahead over a three to four year time span, with indexes under place, sponsoring organization and subject field.

In the world at large it is becoming more and more convenient to consult on-line databases for forthcoming events, because they can offer up-to-dateness and a comprehensive coverage. A good example is *Meeting agenda*, which lists forthcoming meetings, congresses, conferences, exhibitions, colloquia, and seminars up to three years in advance, providing world coverage in the technical, scientific and social science sectors. The host is Questel.

Once meetings and conferences have taken place there is the task of verifying if and when the proceedings have been published and generally made available. By far the best source is the British Library Document Supply Centre's *Index of conference proceedings*: it is the largest bibliography of its kind in the world and lists the proceedings of more than 220 000 conferences in all subjects and languages. A further 18 000 or more are added each year. The *Index* is issued monthly with regular printed cumulations; there is also an 18 year cumulation covering the period 1964–1981 available on microfiche.

To trace a specific conference paper it is necessary to consult the *Conference paper index* (formerly called *Current programs*), published by Cambridge Scientific Abstracts, and covering the life sciences, physical sciences and engineering. Approximately 48 000 individual conference papers are cited annually. The Institute of Scientific Information, Philadelphia, estimates that about 10 000 scientific meetings take place each year (conferences, seminars, symposia, colloquia, conventions and workshops), and that three quarters of them result in a published record. Details are contained in ISIs *Index to scientific and technical proceedings (ISTP)*.

A German-produced guide to conference proceedings is the *Gesamtverzeichnis der Kongresschriften (GKS)*, published by the Deutsches Bibliotheksinstitut, Berlin, and containing details of 70 000 titles.

Conclusions

Of the three types of grey literature discussed above – namely dissertations, translations and meetings papers – the last named cause the most problems. Dissertations, whilst not always readily available, are nevertheless well-indexed; and translations have given rise to specialist centres and indexes which control and co-ordinate their announcement and availability. Meetings papers however, owing to the provisional and sometimes transient nature of their contents, are much more elusive. For this reason the *Directory* compiled by Simonton (see Appendix A) is especially useful in identifying meetings paper series and establishing points of origin if not of availability.

References

Boyer, C.V. (1972) *An analysis of the doctoral dissertation as an information source*. ED–065 157

British Standards Institution (1983) *Specification for presentation of translations*. BS 4755:1971

British Standards Institution (1986) *Recommendations for the presentation of theses*. BS 4821:1972 (1986).

Morton, B. (1983) JPRIS and FBIS translations: polycentrism at the reference desk. *Reference Series Review*, **11**, (1) 99–110

Staiger, D.L. (1973) Separate article distribution as an alternate to journal publications. *IEEE Transactions on Professional Communications*, **PC–16** (3) 107–112, 177

Wood, D.N. (1976) The foreign-language problem facing scientists and technologists in the United Kingdom. *Journal of documentation*, **23** (2) 117–130

Wright, K. (1987) *Translations of scientific and technical literature – a guide to their location*. N88–23686/4/GAR

CHAPTER SIX

The importance of microforms

Introduction

As the volume of grey literature, particularly reports, grew it became apparent that there was a need for a method of producing documents which offered convenient storage, cheapness of reproduction and ease of transmission. The solution lay in various types of microforms. Initially reports were issued on microfilms and microcards. At first 16 mm and later 35 mm microfilm was used to supply early American reports to Europe. In most countries on the receiving end the film was cut into its separate report lengths and then stored in small canisters marked for retrieval and subsequent reading and printing. Microcards too were commonly used for reports literature, especially the 5 x 3 inch size; since they were opaque, however, microcards presented special technical difficulties with regard to enlarging and duplicating, although they could be read easily enough given a suitable reader.

Microfiche has now ousted microcard and microfilm as the prime medium for reports and other separately issued documents converted to microform. Microfiche are sheets of microfilm containing multiple microimages in a grid pattern and a title which can be read without magnification. The term microfiche seems to have entered the English language in the 1950s and derives from the French *fiche*, a sheet or card. The plural can be fiche or fiches. Williams (1970), in reviewing papers on the early development of microfiche in France and Germany just prior to World War II, comments that it would be rash to speak of the 'inventor' of the microfiche, since it is one of the oldest microforms. Microfiche are available in two forms: micronegatives – that is, a clear image on a black background, which prints to give a conventional black image on white paper; and micropositives – that is, a black image on a clear background. Some microfiche combine the two forms on one sheet – for example individual papers in the *SAE Transactions* series – but generally microfiche are made available as micronegatives. It is worth noting that in 1969 the United States Defense Documentation center (DDC) (now called the Defense Technology Information Center – DTIC) established an experimental policy allowing DDC users to order microfiche in either positive or negative film, since a considerable number of customers had maintained that positive microfiche were more legible on viewing equipment. After a couple of years it became evident that customer interest in positive microfiche did not justify the expense of the service. DDC also made numerous tests in colour processing, but at present colour microfiche are considered too expensive for large-scale, closely budgeted systems.

The extent to which microforms have been accepted is indicated by figures relating to the stock of the British Library Document Supply Centre in 1988:

Reports in microform	3 010 000 (holdings)
	130 000 (annual intake)
Microfilm	Roll micro film – over 1700 miles
Microfiche	(Other than reports) 300 000 items

Such numbers confirm the findings of a study by Salisbury (1965) that microfiche offer the following advantages:

(1) accessibility — searches are localized;
(2) economy — less storage space is required;
(3) clarity — a superior image quality;
(4) speed — production and distribution are quicker;
(5) durability — microfiche have a long life;
(6) security — less administration is needed.

The use of microfiche for scientific and technical reports has also been examined by Williams and Broadhurst (1975).

Standardization

The path towards a standard microfiche format has been a tortuous one. When in the early 1960s microfiche began to be issued on a large scale, two standard sizes were employed, NASAs 5 x 8 inch and USAECs 3 x 5 inch. Then in 1964 the Office of Technical Services, hitherto a user of 35 mm microfilm, turned to 4 x 6 inch (104 x 148.75 mm) microfiche, a format endorsed by

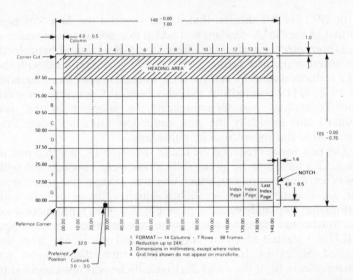

Figure 6.1 **Standard microfiche format** (with acknowledgements to NMA)

COSATI. This particular size was capable of carrying 72 frames of 6 rows of 12, with a reduction of 20:1. In practice however the first row of the fiche was reserved for eye-legible details, and only on trailer fiche did the six rows carry microimages. Thus the term '60 Frame format' was preferred. The other major reports issuing agencies quickly followed suit, and for many years afterwards microfiche were usually of this size. The development influenced other organizations inside and outside the United States, who likewise adopted the microfiche as a dissemination medium for reports. The National Micrographics Association (NMA) of the United States (formerly the National Microfilm Association and now part of the Association for Information and Image Management) has always played an important role in standardization, and the first NMA specification for microfiche, designated M–1 was issued in 1963, and revised in 1967. The 1972 version eliminated the 20:1 reduction, 60 – frame format, 105 x 148 mm microfiche, and now provides for only one standard size with one format, the 105 x 148 mm size with 98 frames at a 24:1 reduction. This size and format have become worldwide standards (see *Figure 6.1*).

A series of British Standards is available based on drafts prepared by the Photographic Standards Committee, and covering formats for 60 and 98 frames; formats for 208, 270, 325 and 420 frames; and computer output on microfiche (COM), A6 size (British Standards Institution, 1978, 81, 85).

In 1972 NTIS indicated that it had adopted the 24 x 98 image format of the NMA standard and added that documents announced in *Government reports announcements* (GRA) would henceforth be issued in the new format. AEC and NASA continued with the 20 x reduction ratio format for their documents for the remainder of 1972. NTIS pointed out that the Federal Council for Science and Technology had recommended the adoption of the new format, and gradually the new reduction ratio has come into world-wide use for microfiche publishing.

The new format reduced the image size from 11.75 x 16.5 mm to 10 x 12.5 mm, but no great difficulties were encountered with users' existing equipment, which was able to cope and gave rise to no great outcry about a slightly smaller image. Users with automatic step-and-repeat equipment for the reproduction of microfiche did have to make some modifications. Boston Spa (BLDSC) started receiving NMA standard microfiche from the United States in July 1972; at that time the demand for enlargements from the older 20:1 microfiche was not expected to decline significantly for some years, and so the library installed an automatic enlarger printer specially built to handle the new NMA standard microfiche.

Today, the copying equipment at Boston Spa consists of:

(1) 90 photocopying machines;
(2) 10 microfilm cameras;
(3) 2 microfilm processors;
(4) 4 Tameran Autoprint 200 roll film enlargers;
(5) 2 microfiche/film enlargers;
(6) Several specialist copying machines, including two producing microfiche duplicates and one producing microfilm duplicates.

Economic advantages

The economic advantages of microfiche may be summarized as lower initial costs, cheaper copies, low postage charges, low storage costs and low-cost copying. A NATO analysis (Vessay, 1970), taking into account the cost of making a microfiche, the cost of a copy microfiche, the selling price of a copy microfiche, the cost of a paper copy from a microfiche and postage charges for despatching up to 25 separate microfiche, found that compared with the cost of a document of 50 pages in hard copy, plus postage charges, the savings in the production and distribution of 1000 copies of a 50 page report were, not surprisingly, overwhelmingly in favour of microfiche, provided that recipients did not insist on making further paper copies from the copy microfiche, when the savings quickly disappeared.

Microfiche offer large economies in storage space: a simple index drawer of a purpose-designed microfiche storage cabinet is capable of storing between 1000 and 1500 microfiche – that is, over 100 000 pages of hard copy equivalent assuming some of the frames are not used. The British Library Document Supply Centre claims to have had more extensive experience of handling library materials in microform in a wider variety of formats than any other organization in the United Kingdom and has a wealth of information on the economics of microforms as a storage medium in libraries.

In theory even greater economies of storage can be achieved with a format called ultrafiche – that is with images at a reduction rate of more than 98x. A number of systems were tried on a commercial basis, including that of NCR which produced 105 x 148 mm fiche with images reduced to either 120x or 150x. At the 120x reduction the fiche contained 70 columns and 30 rows, and provided space for 2100 A4 documents. At the 150x reduction ratio, the capacity was even greater. Perhaps the ultimate in ultrafiche was the publication by NCR of the Holy Bible (1245 pages) on a single fiche measuring 50 x 50 mm. In practice however the reports issuing agencies have preferred to stay with the 98x format.

Reader resistance

Reader resistance to microfiche is high, despite the development and refinement of reading machines. Two major problems are the user's inability to move rapidly from one section of a report to another and back again, and place book markers at key points; and the inability to compare two or more microfiche documents simultaneously unless extra viewing machines are situated side by side. Resistance also increased with the number of fiche per original document. Whereas a short paper or report which fits comfortably on one fiche is just about acceptable, a long document running to several hundred pages and requiring a number of trailer fiche certainly is not. The microfiche is however able to offer an advantage of its own in that group viewing is possible. Several people can study the same fiche together by using a reader capable of projecting the image on to a large screen.

It is essential that microfiche are of good optical quality, both from the user's point of view for ease of reading and from the reproduction angle, since a good microfiche copy is vital before a satisfactory print can be taken from it by an automatic enlarger printer. Some defects such as out-of-focus frames and misaligned pages, are due to poor quality control at the filming source. Others derive from the poor condition of the original document and many reports issuing agencies are still obliged to put out warning notices such as 'This document has been reproduced from the best available copy furnished by the sponsoring agency. Although it is recognized that certain portions are illegible, it is being released in the interest of making available as much information as possible'. Other notices in use include: 'Reproduced from the best available copy' and 'Copy available does not permit fully legible reproduction'. Such warnings do little to assuage the outraged feelings of the indignant recipient, who is at a loss to know how, in the age of the total quality organization, such substandard originals can be allowed to get into the system. One solution lies in the stricter enforcement of the format standards discussed in Chapter 4, and in adherence to the recommendations for the preparation of research and development reports and technical manuals for microcopying issued by the British Standards Institution (1977). Further consideration of the technical quality of microfiche reports is contained in the study by Horder (1977).

A further factor which militated against user acceptance of microfiche is the necessity of having to use a machine at all. Many users are deterred simply by the prospect of being required to walk to and sit down in a library or specially designated office to consult a fiche on a viewer. Again, many people in a wide variety of disciplines and professions do a great deal of their employment-associated reading whilst travelling by train, or plane, or in the back of a limousine. In such circumstances a portable viewer would be of great benefit, but there is little evidence of the acceptance and use of such devices.

Many authorities are firmly of the opinion that the most important factor acting against user acceptance is the limited availability of viewers, and the view is that if the position could be improved in this respect, other means of encouragement could then be used.

A more optimistic note was struck in a study (Christ, 1972) of the Libraries and Information Systems Center of Bell Telephone Laboratories, which has a Technical Reports Center with a primary responsibility for the acquisition, announcement and distribution of externally generated reports to the company's scientists and engineers. Reports are selected from various abstract journals and then listed in a semi-monthly internal publication called *Current technical reports* (CTR). The normal response to requests for reports announced in CTR is by hard copy loans, but it was decided to use technical reports as a vehicle for evaluating the microfiche format. The test ran for some 9 months, during which time requests were fulfilled by microfiche, and at the end of it one of the main conclusions was that the reading habits of the test group had not been adversely affected by the use of this medium. In fact the data, comments and observations gathered during the test suggested that in general, Bell Laboratories technical staff would accept and use microfiche, albeit grudgingly, provided care was taken to design the distribution system to their needs and preferences.

Readers and reader-printers

Since satisfactory equipment is such a key factor it is worth considering some of the technical aspects of readers and reader-printers. Many different types are currently available and the user has a very wide choice – to mention some manufacturers and not others would clearly be unfair – and a decision is best made as to which is the most suitable after consulting buyers' guides, checking independent assessments, obtaining quotations and visiting trade exhibitions.

The list of standards relating to microforms in general and microfiche in particular, both British and foreign, is very extensive, as is the literature on the availability and evaluation of equipment and methods. Impartial advice is available in the United Kingdom from Cimtech, the British National Centre for Information Media and Technology, founded in 1967 as the National Reprographic Centre for documentation (NRCd). Cimtech specializes in document imaging, micrographics, records management, publishing systems and related subjects; news about its activities can be found in its journal *Information media and technology*.

The basic requirements in any microform reader are a sharp image and a sufficiently high light intensity, features which can quickly be determined by a short test in normal room lighting, but with no bright lights or sunlight shining directly on to the screen. Important characteristics to look for in addition to purely optical ones are facilities for indexing line and frame numbers, and the ability to rotate the image to view diagrams, drawings and tables printed sideways in the original.

Reader-printers are similar in principle to machines designed solely for viewing, but permit enlarged format printing of individual frames, usually as A4 sheets. Since they normally operate at a relatively slow speed (of the order of half a minute per page), any enlarging is best confined to selected sections of the fiche, and the full-size reproduction of entire documents should be entrusted to centres such as Boston Spa or to local copying bureaux operated by the manufacturers and suppliers of reprographic equipment.

The actual making of microfiche from the original hard copy documents requires skilled operators and expensive cameras, and again is undoubtedly best left to specialized agencies. Similarly the duplication of microfiche involves the employment of very efficient but fairly expensive equipment which requires a high throughput to justify its cost, and again is best left to central agencies, although it is possible to purchase or lease microfiche duplicators for use in ordinary office conditions.

Conclusions

Although by definition the final product is small, in some applications very small, the world of microforms is a large one and the end of development is by no means in sight. A good measure of standardization has been achieved and this has resulted in stability which in time has led to the acceptance of microforms as not quite the necessary evil they once were. In particular microfiche have been used in applications ranging from spare parts lists in garages to supplements to catalogues in large libraries, and of course the invaluable *British books in print* (BBIP), now rechristened *Whitaker's books in print*. This combination of diversity and familiarity has meant a greater acquiescence to the medium, but most users will still point out that the only acceptable reading aid is a pair of spectacles.

References

British Standards Institution (1977) *Recommendations for preparation of copy for microcopying*. British Standard 5444:1977

British Standards Institution (1978) *Microfiche formats of 208, 270, 325 and 420 frames (except COM)*. BS 4187, part 3:1978

British Standards Institution (1981) *Microfiche: specification for formats for 60 and 98 frames*. BS 4187, part 1:1981 British Standards Institution (1985) *Specification for computer output microfiche (COM), A6 size*. BS 5644:1978 (1985)

Christ, C.W. (1972) Microfiche: A study of user attitudes and reading habits. *Journal of the American Society for Information Science*, **23** (1) 30–35

Gabriel, M.R. (1978) *Micrographics 1900–1977 – a bibliography*. Mankato, Minnesota: Minnesota Scholarly Press

Hernon, P. (1981) *Microforms and government information*. Westport, Connecticut: Meckler Corporation

Horder, A. (1977) *Technical quality of microfiche reports: a preliminary study*. Hatfield, Herts: National Reprographic Centre for Documentation

J Whitaker and Sons Ltd (1982) Microfiche readers – a survey *Bookseller* (4018) 2278–2279 (1982)

Martin, N.J. and Zink, S.D. (1984) (Eds.) *Government documents and microforms: management issues.* Proceedings of the 4th annual government documents conference and the 9th annual microforms conference. Westport, Connecticut: Meckler Corporation

Salisbury, J.T. (1965) *A study of the application of microfilming to the production distribution use and retrieval of technical reports.* Report AD–615 800

Vessey, H.R. (1970) *The use of microfiche for scientific and technical reports.* AGARD Advisory Report 27, N70–39851

Williams, B.J.S. (1970) *Miniaturised communications: review of microforms.* London: Library Association

Williams, B.J.S. and Broadhurst, R.N. (1975) *Use of microfiches for scientific and technical reports: considerations for the small user.* Hatfield, Herts: National Reprographic Centre for Documentation

CHAPTER SEVEN

Aerospace

Introduction

The opinion is often advanced that a progressive industrial society needs a pioneering spearhead technology to stimulate growth, promote innovation and set standards of excellence for a nation's entire range of manufacturing and production activities. In many countries aerospace fulfils this role, even though dissenting voices are raised from time to time at the colossal expenditures involved. The European Space Agency presents its summaries for income and expenditure in MAUs and KAUs (Millions of Accounting Units and Thousands of Accounting Units), and notes in its annual reports that 'the adjustment of contributions due to conversion rate variations gave rise to animated discussions among the Member States'.

Factors which help to account for the large amount of grey literature, particularly reports literature, in the aerospace sector are directly related to this considerable expenditure. Among them are:

(1) The immenseness of the projects involved (for example the project proposed in the Ride report (Ride, 1987) for a lunar-launched manned mission to Mars is likely to require a NASA budget of $30 billion a year by the year 2000);
(2) The employment of large numbers of highly qualified, highly skilled personnel;
(3) The importance of aerospace in terms of national security;
(4) The need to conduct research and development programmes on a broad front simply to maintain a strategic advantage or a competitive position; and, not least,

(5) The ruling that contractors working on government projects are required to report on progress at regular intervals.

Security has always been a great feature of aerospace information, and many publications stay inaccessible to those without a need to know. Even if this classified material is disregarded, there still remains a large body of aerospace documentation which is actively promoted, firstly to benefit the aerospace industry in general by making available a common core of knowledge, and secondly to try and justify and if possible recoup some of the costs involved by promoting technology transfer and product spin-off.

Reports issuing agencies

Many agencies throughout the world issue and co-ordinate reports on various aspects of aerospace, and only representative bodies for some of the major countries actively concerned can be indicated here. For a comprehensive list of agencies reporting in any one year the reader should consult the annual cumulative source index issue to *Scientific and technical aerospace reports* (STAR), discussed below, where entries are arranged alphabetically by the organization responsible for a document's original appearance.

Aeronautical Research Council

In Great Britain the principal agency with a major output of reports was the Aeronautical Research Council, a body which under one name or another goes back to 1909 when the Advisory Committee for Aeronautics (with functions virtually identical with those of the eventual ARC) was appointed to advise the Prime Minister of the day. In 1920 it became the Aeronautical Research Committee, and in 1945 the Aeronautical Research Council until its disbandment in 1980.

The Council's function was purely advisory – that is it had no executive power over the conduct of research and possessed no funds of its own, and was concerned purely with research, as distinct from development. ARC published through HMSO two series of reports embodying the results of research. The manuscripts were initially submitted to ARC by government research establishments, firms and universities for discussion and consideration as to their suitability for publication. The two series were the *Reports and memoranda series* (R & M), reports having permanent value and printed by a photosetting process; and the *Current papers series* (CP), of ephemeral interest but occasionally used to secure speedy publication of data or information of immediate importance, duplicated by reprographic means. HMSO published a sectional list (no. 8) devoted solely to ARC and between them the two ARC series constituted a major communications channel for the results of aeronautical research in the United Kingdom.

ARCs system of report identification could sometimes cause problems, because although ARC documents originated in many different organizations, these organizations were not named in the HMSO catalogues and in the Sectional List. Thus the reader had no means of ascertaining the original number assigned by the issuing agency. This information could be gleaned from other sources, but the treatment varied according to which announcement journal was consulted. For example, ARC CP 1173 was listed in HMSO Sectional List no 8 simply as 'A parallel motion creep extensometer' by J.N. Webb; in *R and D abstracts* the corporate author was given as the Aeronautical Research Council, Teddington with an ARC internal reference number (ARC 32319); whereas in STAR the corporate author was the Royal Aircraft Establishment (RAE) Farnborough, Structures Department (N72–28475), with the RAEs original reference RAE–TR–70068 added for good measure. The first words on the cover of the document itself were 'Ministry of Defence (Aviation Supply)'. The Royal Aircraft Establishment itself, with its world-wide reputation for technical excellence and painstaking thoroughness, is a major source of reports, many of which were reissued as ARC documents; other RAE reports are not generally available because they are security classified. Enquiries concerning ARC documents should now be addressed to RAE.

A tidying up operation used to take place in that all ARCs *Reports and memoranda* were eventually bound into *Annual technical reports volumes*, and as far a possible the bound volume for a given year contains the R & Ms whose original report dates correspond with that year, but there are exceptions.

Other United Kingdom bodies which regularly publish reports on aeronautical topics are government research stations such as the National Gas Turbine Establishment (NGTE). In the academic sphere a number of institutions are noted for their report series, particularly the Department of Aeronautics and Astronautics at Southampton University, and Cranfield Institute of Technology, (previously known as Cranfield College of Aeronautics). A more detailed account of the aerospace engineering scene is given by MacAdam (1985).

NASA

In the United States an organization of similar age to ARC but very much larger and more powerful is the National Aeronautics and Space Administration (NASA), which was created in 1958 out of the National Advisory Committee for Aeronautics as a civilian agency with the task of accomplishing the American aeronautics and space programmes. With regard to information, the National Aeronautics and Space Act of 1958 required that 'the aeronautical and space activities of the United States be so conducted as to contribute to the expansion of human knowledge of phenomena in the atmosphere and space. The Administration shall provide for the widest practicable and appropriate dissemination of information concerning its activities and the results thereof'. It is greatly to the credit of NASA that these unequivocal words have been interpreted in the spirit as well as the letter, with the result that a very fine information service has developed, and to apply the term grey literature to its output seems singularly inappropriate when the acronym for its announcement service STAR implies brightness.

Two offices of NASA have a major responsibility for information services – these are the Public Information Division of the Office of Public Affairs, which handles non-technical information, especially for the news media; and the Scientific and Technical Information Division of the Office of Technology Utilization, whose responsibilities include science and technology and world-wide aerospace information. In addition to the announcement journal *Scientific and technical aerospace reports* (STAR), discussed below, NASA disseminates several formal publications series some of the most important of which are as follows:

Contractor reports (NASA–CR–). Technical information generated in the course of a NASA contract and released under NASA auspices;

Technical memoranda (NASA TM–X). Information which receives limited distribution because of its preliminary or classified nature;

Technical notes (NASA TN–D–). Information of lesser scope, but still important as a contribution to knowledge;

Technical reports (NASA TR–R). Scientific and technical information considered important, complete and a lasting contribution to existing knowledge;

Technical translations (NASA–TT–F). Information originally published in a foreign language, but considered sufficiently valuable to NASAs work to merit distribution in English.

Other series of a more general kind are:

Conference publications (CP) – records of the proceedings of scientific and technical symposia and other professional meetings sponsored or cosponsored by NASA.

Research publications (RP) – compilations of scientific and technical data deemed to be of continuing reference value.

Special publications (NASA SP–). Special publications are often concerned with subjects of substantial public interest; they report scientific and technical information derived from NASA programmes and are for audiences of diverse technical backgrounds. An example of a Special Publication aimed at a general audience is *Records of achievement* (NASA–SP–470) published on the occasion of NASAs 25th anniversary, and including an illustrated narrative on the moon landings and the Voyager journeys.

Technical papers (TP) present the results of significant research conducted by NASA scientists and engineers.

A Catalog of special publications . . . (NASA–SP–7063) covering the years 1977–1986 and citing 2311 references is available free of charge. In addition, any user of the NASA system will benefit from studying the publication *The NASA Scientific and technical information system . . . and how to use it*, also available free of charge.

As noted in Chapter 1, the predecessor of NASA was the National Advisory Committee for Aeronautics (NACA), and details of the work carried out under its auspices, a great deal of which is of a fundamental nature and consequently still in demand, can be obtained by consulting the *Index of NACA technical publications 1915–1949* (1950), and its supplements up to 1960. For a history of NACA and NASA, see N82–14955, *Orders of magnitude*, covering the period 1915–1980.

European agencies

In continental Europe several agencies are active in the dissemination of aerospace information. Firstly there is the internationally composed Advisory Group for Aerospace Research and Development (AGARD), the mission of which is to bring together the leading personalities of the NATO nations in the fields of science and technology relating to aerospace. AGARD distributes unclassified reports and other publications to NATO member nations through national distribution centres. The national distribution centre for the United Kingdom is the Defence Research Information Centre, and the UK purchase agency is the British Library Document Supply Centre. A cumulative index to AGARD publications issued since 1952 is regularly updated, and provides information on the major AGARD series:

(1) *Advisory reports*, AR;

(2) *Reports* RP;
(3) *Agardographs* AG;
(4) *Conference proceedings*.

Full bibliographical references and abstracts of AGARD publications are given in *Scientific and technical aerospace reports* and *Government reports announcements and index*. See for example N87–29369, abstracts and indexes for AGARD publications published during the period 1983 to 1985, and prepared with the help of the NASA database.

In France, where AGARD has its headquarters, the French national agency concerned with aerospace is the Office National d'Etudes et de Recherches Aérospatiales (ONERA), which has published several series of reports and technical notes since 1947. In Germany a similar organization responsible for various series of aerospace publications is the Deutsche Forschungs– und Versuchsanstalt für Luft- und Raumfahrt (DFVLR), the German Aerospace Research Establishment. DFVLR is the largest research establishment for engineering sciences in the Federal Republic and publishes reports in two major series:

(1) *DFVLR Forschungsberichte* (FB series);
(2) *DFVLR Mitteilungen* (Mitt. series).

An index is published annually. Most of the FB and Mitt. documents are translated into English by the European Space Agency and published as *ESA Technical translations* (ESA TT). In addition to its generally available publications, DFVLR issues about 1300 internal scientific reports annually.

In Europe as a whole grey literature in the aerospace field is co-ordinated by the European Space Agency (ESA), which was formed out of two earlier bodies, the European Space Research Organization (ESRO) and the European Organisation for the Development and Construction of Space Vehicle Launchers (ELDO). The member states are Austria, Belgium, Denmark, France, Germany, Ireland, Italy, Netherlands, Norway, Spain, Sweden, Switzerland and the United Kingdom. Finland is an associate member and Canada a co-operating state. The purpose of the Agency is to provide for exclusively peaceful purposes co-operation among European States in space research and technology. ESA has an extensive publications programme with various series of reports with categories similar to those issued by NASA, that is to say conference proceedings, technical reports, technical memoranda and so on. In 1987 ESA processed over 4000 items for input to *Scientific and technical aerospace reports*. The Agency also operates a very important information retrieval service, ESA/IRS, which offers on-line access to over one hundred databases in all subjects and disciplines, not just aerospace. Access in the United Kingdom is through the ESA/IRS National Centre, the Department of Trade and Industry. Developments in ESA can be followed through the quarterly *ESA Bulletin*.

Announcement services

STAR

The aerospace industry has always been well-organized in its handling of information, and it possesses a major abstracting and indexing journal covering current world-wide grey literature on the science and technology of space and aeronautics, namely *Scientific and technical aerospace reports* (STAR). Previous guises of STAR were *NASA Technical publications announcements* and the *National Advisory Committee for Aeronautics research abstracts*. Publications abstracted in STAR cover a large section of grey literature documents and include scientific and technical reports issued by NASA and its contractors, other US government agencies, corporations, universities and research organizations throughout the world. Pertinent theses, translations, NASA owned patents and patent applications are also abstracted.

The subject scope of STAR includes all aspects of aeronautics and space research and development, supporting basic and applied research, and applications. Aerospace aspects of earth resources, energy development, conservation, oceanography, environmental protection, urban transportation and other topics considered of high national priority in the United States are also covered. STAR is arranged in ten major subdivisions divided into 76 specific subject categories and one general category/division. The main subject divisions are:

(1) Aeronautics;
(2) Astronautics;
(3) Chemistry and materials;
(4) Engineering;
(5) Geosciences;
(6) Life sciences;
(7) Mathematics and computer sciences;
(8) Physics;
(9) Social sciences;
(10) Space sciences.

Despite the careful specification of NASAs subject interests, it has often been noted that STAR is full of surprises. Thus for example it is possible to find details of a thesis on weekend cottage recreation in Bavaria (N70–26211). Again, NASAs search for truth is not confined entirely to outer space – in the 1970s it participated in the Shroud of Turin Research Project (STURP), using a system of analysis adopted from the Apollo-programme.

Entries in STAR are arranged in an unbroken sequence of accession numbers – for example N88–10321, where N stands for NASA and the digits for the year of accession and the accession number, see *Figure 7.1*. Although the accession number sequence is unbroken, it is not uninterrupted, owing to the special analytic treatment given to certain publications such as conference reports or book translations which have separate abstracts written for their component papers or chapters. Citations and abstracts of the components follow immediately after those of the parent document. When the subject content of a component differs sufficiently from the general content of the parent, a cross-reference citation is also printed at the end of the category to which it would normally have been assigned if it were independent.

Five indexes are included in each issue of STAR

(1) Subject index;
(2) Personal author index;
(3) Corporate source index;
(4) Contract number index;
(5) Report/accession number index.

Two of these indexes are of special significance.

Firstly, as from volume 12, 1974, STAR introduced a change with respect to the subject index. Previously the Notation of Content (NOC) rather than the title of a document had been used to provide a more exact description of the subject matter. The NOCs were arranged under each broad subject heading in ascending accession number order. Under the new arrangement NOCs no longer appear in STARs subject index or in other current NASA information products. Instead titles of the referenced documents are used with a synthesized title extension added in those cases in which the title is insufficiently descriptive. The object, according to NASA, is to enhance the utility and relevance of the subject index.

Secondly the report/accession number index is of particular importance because of NASAs commendable practice of giving full details of a report's original number, that is the one assigned by the issuing agency responsible for the work reported. Report numbers may thus be in the main NASA series, as for example the

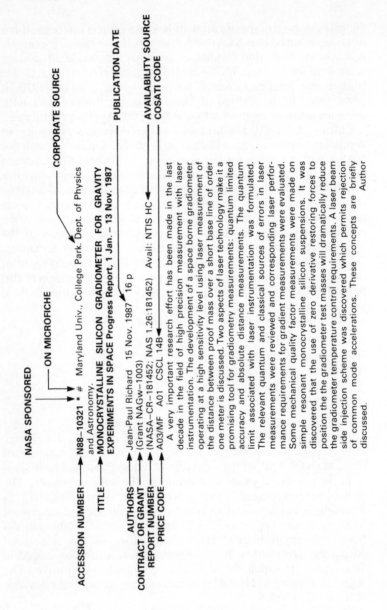

NASA SPONSORED

ON MICROFICHE

CORPORATE SOURCE

PUBLICATION DATE

AVAILABILITY SOURCE
COSATI CODE

ACCESSION NUMBER → **N88–10321** * # Maryland Univ., College Park. Dept. of Physics
and Astronomy.
TITLE → **MONOCRYSTALLINE SILICON GRADIOMETER FOR GRAVITY
EXPERIMENTS IN SPACE** Progress Report, 1 Jan. – 13 Nov. 1987

AUTHORS → Jean-Paul Richard 15 Nov. 1987 16 p
CONTRACT OR GRANT → (Grant NAGw–1003)
REPORT NUMBER → (NASA–CR–181452; NAS 1.26:181452) Avail: NTIS HC
PRICE CODE → A03/MF A01 CSCL 14B

A very important research effort has been made in the last decade in the field of high precision measurement with laser instrumentation. The development of a space borne gradiometer operating at a high sensitivity level using laser measurement of the distance between proof mass over a short base line of order one meter is discussed. Two aspects of laser technology make it a promising tool for gradiometry measurements: quantum limited accuracy and absolute distance measurements. The quantum limit associated with laser instrumentation was formulated. The relevant quantum and classical sources of errors in laser measurements were reviewed and corresponding laser performance requirements for gradient measurements were evaluated. Some mechanical quality factor measurements were made on simple resonant monocrystalline silicon suspensions. It was discovered that the use of zero derivative restoring forces to position the the gradiometer test masses will dramatically reduce the gradiometer temperature control requirements. A laser beam side injection scheme was discovered which permits rejection of common mode accelerations. These concepts are briefly discussed.   Author

Figure 7.1 **Typical citation and abstract in STAR** (with acknowledgements to NASA)

Contractor Report (CR) series; in series used by NASAs own research centres such as the E series of the Lewis Research Center; or in series used by other organizations both in America and overseas.

The last named category, as noted above, includes reports from RAE. The practical result is that no matter by what number an aerospace document is cited and sought, it is possible to check against the accession number N sequence and so determine its availability. Occasions when such citations occur are typically in personal contacts among scientists and engineers of various establishments and companies, who exchange details of the documents issued by their respective organizations without reference to the NASA system. Subsequently, months or even years later, a library or information department may be asked to supply one of the documents so mentioned, and often the quickest and surest way is to check the N sequence. The report/accession index provides the means for doing this.

Cumulative index volumes are published semi-annually and annually. The introductory remarks to each issue of STAR are particularly well detailed, covering the method of announcement of NASA publications in STAR, the distribution of NASA publications, eligibility and registration for NASA information, products and services, the public availability of documents announced in STAR (that is: in the United States), the European availability of NASA and AGARD documents, and the indexing vocabulary. The introduction also includes examples of typical citations and abstracts and, always provided the points raised are observed, the user will have no difficulty in tracing and obtaining the documents required.

With earlier issues of STAR, up to and including 1971, a sixth index was necessary to relate accession numbers to report numbers since entries were not arranged in one unbroken sequence of N numbers.

The twofold advantage of STAR, as stated by NASA and borne out by the experience of many users over the years, is a current awareness service achieved through frequency of publication (24 issues per year) and a retrospective searching tool arising from thorough indexing in each issue and twice-yearly cumulative indexes. In fact it is difficult to overstate the value of STAR to any organization intimately, or only peripherally concerned with aerospace.

With regard to NASA accession numbers, readers will benefit greatly by familiarizing themselves with one or two limitations concerning the use of this device to secure copies of certain documents. The NASA accession number is sufficient when ordering NASA or NASA-sponsored documents, marked with an asterisk (*) from NTIS. When non-NASA documents are wanted (no asterisk) either direct from the issuing agency or from other sources, especially NTIS, it is vital that additional background information, including the report number assigned by the originating agency be given. A further symbol commonly used is #, indicating that a document is available on microfiche. Documents which cannot be microfiched, but for which a one-to-one facsimile can be provided, are marked with a plus sign (+) instead of #. Occasionally reports entries carry no asterisk (i.e. not NASA or NASA-sponsored) no # (i.e. no microfiche), no plus sign (i.e. no one-to-one facsimile), and availability then depends on the issuing agency's willingness to supply. Documents so listed are a result of NASAs policy of making STAR as comprehensive as possible.

Some reports, for various reasons, may not be available direct from their originators, but a different version of the same document may be available from another source, quite often on microfiche only. Chillag (1973) quotes the example of N71–17390, which was originally prepared as report NGTE NT766, but since the paper was intended for publication, it was not released in that form for general use. It finally found a home in a volume of Agard *Conference proceedings*, AGARD–CP–73–71 (N71–17372). Such literature is not so much grey as labyrinthine in nature!

The British Library Document Supply Centre maintains an extensive collection of NASA and NASA sponsored publications, but some documents (marked * and/or #) are not always available on demand, even though they have been issued for some time – usually the British Library asks the reader to reapply in a few weeks' time, and the interval is used to secure a copy from the United States.

STAR has a companion publication, *Limited scientific and technical aerospace reports* (LSTAR), which announces reports that are security classified, and unclassified reports with availability limitations.

International aerospace abstracts

STAR is published on the 8th and 23rd of each month and is complemented by *International aerospace abstracts* (IAA), which appears on the 1st and 15th of each month and is an abstracting and indexing journal providing world-wide coverage of the open aerospace literature, including science and trade journals, books and meetings papers. Bibliographical citations and abstracts are arranged in the same subject categories as those used in STAR. Each entry is prefixed by an *IAA* accession number, and each issue contains a subject, personal author, contract number and accession number index. An index is also provided to meetings papers – for example those published by ASME – and surprisingly to reports, although very few are included, since this is the stated domain of STAR.

Index aeronauticus

An abstracting journal published in Britain for over 20 years (the last issue appeared in 1968) and now only of historical interest was *Index aeronauticus*, compiled by the staff of the Ministry of Technology and its predecessors, whose primary purpose was to draw attention to articles of interest in scientific and technical journals, and to published papers and reports within the field of aeronautics. *Index aeronauticus* deserves a mention in any survey of aerospace information sources because of two striking characteristics: its dumpy A5 format, and its valiant use of the Universal Decimal Classification to arrange its abstracts.

Retrieval systems

As the reports and other literature amassed by NASA continued to grow, and as the operational need to exploit it to the full became more urgent in the successful pursuance of NASAs programmes, an information retrieval system known as NASA/ RECON (REmote CONsole) was developed and expanded into a national and eventually an international network.

The information retrieval language used was a system known as DIALOG, developed by Lockheed and described by Summit (1967), which allowed an important and now familiar facility, the direct interaction between the search requester and the computer. It provided means, via a display console, for the user to determine: (1) what terms were alphabetically related to a specific search term or set of terms; (2) the number of documents in the collection indexed under each displayed term; (3) terms which were conceptually related to any given term; and (4) what documents dealt with any term or set of terms chosen for searching.

The Lockheed system was one of the first to offer a large-scale online search service, and now makes available over one hundred databases, of which about forty per cent are unique to DIALOG.

DIALOG acts as a host for the *Aerospace database*, which is the on-line version of *Scientific and technical aerospace reports* and *International aerospace abstracts*. Summaries are provided for over ninety per cent of the records from 1972 onwards. However, as is pointed out in the introduction to each issue of STAR, only certain organizations are eligible to register for NASA products and services, and in consequence, access to, use of, or distribution of data contained in the Aerospace Database is limited to users within the United States and is intended only for organizations or persons whose primary activities are within the United States or its territories. Arrangements for access by overseas users are subject to the provisions of a scientific and technical information agreement between NASA and a central government agency in the country concerned. In the case of the United Kingdom, the agency is the Department of Trade and Industry, through its IRS–DIALTECH service. DTI points out that unlike most other database producers, NASA does not charge a royalty for the use of its file. Instead it restricts access to organizations working in aerospace and related areas who can provide an input to the databases. The input NASA requires is, over a period of time, one appropriate technical report per hour of access. In order to use NASA it is necessary to sign an Agreement (labelled ESA–Direct), and complete a Statement of Justification for access. DTI will supply without charge a copy of NASA SP–7200 *NASA Guidelines on report literature*, which describes the types of documents which NASA requires for inclusion in its database. Full details on how to comply with NASAs requirements are available from the DTI, and a description of the RECON system has been provided by Jack (1982).

Technology transfer

Nowadays the virtues of technology transfer are widely recognized, and organizations abound which preach its advantages to industry with missionary zeal. In all this NASA played a pioneering role, for it was recognized that ideas initially developed at great public expense to meet the exacting requirements of the aerospace industry were often capable of modification for application in a wide range of everyday areas of industry. Consequently, as a special aid to persons and companies not directly involved in aerospace activities the NASA Office of Technology Utilization began to issue numerous publications regarding advances likely to be capable of transfer from one field to another, and examples of the successful application of 'spin-off' technology began to appear. Moreover the term technology was interpreted in a very wide sense, and not confined solely to manufacturing and processing industries. Examples of technology transfer have been reported in various biomedical projects and in public sector initiatives concerned with air and water pollution, housing and urban constructions, law enforcement and crime prevention, transport and fire safety. Many individual cases of technology transfer have been reported by NASA – for example a fish farm in Texas profited from data on water recycling and purification systems, originally obtained by NASA in pursuit of its own requirements in the space vehicle programmes (see N72–28934).

The main media for announcing possible technology transfer projects are *Tech briefs, Technology utilisation reports* and *Technology utilisation surveys.*

Tech briefs are short announcements of new technology derived from the research and development activities of NASA. They emphasize information likely to be transferable across industrial, regional and disciplinary lines, and are issued to encourage commercial applications.

Technology utilisation reports are of a far more substantial nature than the *Tech briefs* and provide detailed descriptions of developments and innovations of promise. Similarly with *Technology utilisation surveys* which are comprehensive state-of-the-art accounts identifying substantial contributions to technology by NASA researchers or NASA contractors. Both types of document appear in the NASA SP series noted above.

NASA also produces *Research and technology objectives and plans* (RTOPs) which are summarized to facilitate communication and co-ordination among interested technical personnel in govern-ment, industry and the universities (see for example N88–14894). Specific issues are also reported in depth, as for example the use of expert systems in technology transfer (N88–14882).

Technology transfer as an identified activity has spread beyond the aerospace industries. Knox (1973), in a survey of systems for technological information transfer, noted in particular the efforts of the Department of Defense, the Atomic Energy Commission and the Office of Education; and he pointed out that in each case one of the distinguishing features was the emphasis on the technical report. Technology transfer also takes place by osmosis and diffusion and industry has not been slow to exploit opportunities. One of the by-products of the programme which culminated in the first moon landing in 1969 has been the development by a whole range of companies of smaller and more reliable lightweight computers.

Summary

The grey literature of the world of aerospace, as with other subject areas, is characterized by a diversity of form and content.

Hundreds of agencies contribute many thousands of documents each year to a common pool. Consistent and comprehensive efforts are made world-wide to ensure that as many of the publications as possible reach a wide and diffuse readership at the earliest possible opportunity, subject only to the restrictions of security considerations.

References

Chillag, J.P. (1973) Don't be afraid of reports *BLL Review*, **1**, (2) 39–51

Jack, R.F. (1982) The NASA Recon search system. *Online*, **6**, November, 40–54

Knox, W.T. (1973) Systems for technological information transfer *Science*, **181** (4098) 415–419

MacAdam, E.J. (1985) Aerospace engineering in Anthony, L.J. (Ed.) *Information sources in engineering*. Butterworths, London

Ride, S. (1987) *Leadership and America's future in space*. NASA Office of Exploration

Summit, R.K. (1967) Dialog – an operational online reference retrieval system *Proceedings 22nd national conference Association of Computing Machinery*, 51–56

CHAPTER EIGHT

Life sciences

Introduction

The definition of the term life sciences depends on which authority is consulted. A dictionary will say a life science is any of the sciences (such as zoology, bacteriology or sociology) which deal with living organisms; or, such sciences collectively. In the case of grey literature, the life sciences collection of the Dialog Information Retrieval Service offers a convenient way of defining the area under consideration, since it embraces biology, medicine, biochemistry, ecology, microbiology, agriculture and veterinary science. Another approach is that used by the BIOSIS Information System, which says the life sciences relate to the study of living things:
    Animals (inc. Humans) – Plants – Micro-organisms

(1) How they look (anatomy, morphology, cytology);
(2) How they live (physiology, biochemistry);
(3) How they relate to the environment (ecology, behaviour);
(4) How they grow (reproduction, evolution, genetics);
(5) How they are identified (taxonomy, systematics).

For the purposes of this chapter, these various taxonomies have been consolidated into the headings *Medicine and biology* and *Agriculture and food*.

The conventional literature relating to the life sciences has been considered separately and comprehensively in other titles in this series, especially Morton and Godbolt's *Information sources in medical sciences* (1984); Lilley's *Information sources in agriculture and food science* (1981); Wyatt's *Information sources in the life sciences* (1987); and the somewhat older *Use of earth sciences literature*, by Wood (1973).

None of these works highlights reports in particular or grey literature in general as being of special significance, and this is to some extent borne out by the absence of an announcement journal devoted exclusively to grey literature in the life sciences. The title which comes nearest is *Biological abstracts/RRM*, which covers reports, reviews meetings papers, US patents and books. It is the companion publication to *Biological abstracts*, which provides English language abstracts of current research reported in biological and biomedical journals.

An area in the life sciences where grey literature is playing an increasingly important role is the field of environmental problems such as oil spills and river pollution, which do not stop neatly at national boundaries. In a paper by Deschamps (1986), international sources of information on chemicals in the environment are reviewed and discussed with particular attention to the United Nations Environment Programme (UNEP). An important service provided by UNEP is the *International register of potentially toxic chemicals* (IRPTC), the main objectives of which are:

(1) to make the data on chemicals readily available to those who need it;
(2) to locate and draw attention to the major gaps in the available information and encourage research to fill those gaps;
(3) to identify the potential hazards of using chemicals and make people aware of them;
(4) to assemble information on existing policies for control and regulation of hazardous chemicals at national, regional and global level.

The goal of IRPTCs work is the development of a storehouse of information adequate for an understanding of the hazards to health and the environment associated with toxic chemicals.

Information collected by IRPTC can be grouped into two major categories, information on chemicals and information on chemical regulations. Information from conventional and grey literature sources is combined in the production of data profiles for various chemicals, the central activity of the IRPTC staff.

At the Eurpean level, there is the database ECDIN, which has been described by Gilbert (1988), and is the online equivalent of a chemical handbook. ECDIN (*Environmental chemicals data and information network*) is a factual databank created in the framework of the Environmental Research Programme of the Joint Research Centre (JRC) of the Commission of the European Communities at the Ispra (Italy) Establishment. ECDIN deals with the whole spectrum of parameters and properties that might help the user to evaluate real or potential risks linked to the use of a chemical and its economical and ecological impact. ECDIN contains identifications for more than 65 000 compounds, and information is stored in files with a variety of formats. For example, the occupations poisoning file contains concise reports in a narrative style, whilst the symptoms and therapeutic treatment file consists of various checklists.

As with so many other areas of activity which embrace a multitude of scientific and technical disciplines, databases providing environmental information are becoming more and more important, and affect all aspects of the life sciences. A most useful survey of them has been compiled by Breen (1988).

Medicine and biology

On the question of grey literature in medicine and biology, it has been observed in the book by Morton and Godbolt, noted above, that 'the indifference which most scientists feel for the research report literature is due not so much to a low opinion of reports as to a respect for the traditional primacy of the "open" literature'. It has also been noted that as a form of primary communication, reports are of much less significance in medicine and biology than in other areas of science and technology, and a major reason for this state of affairs is that there is far less government controlled industrial research and development than in space projects or nuclear energy activities. Even where state institutions are involved, such as the Medical Research Council and the American Public Health Service, their research reports are published not as communications directly from a research establishment with all the characteristic identifying features of a report, but as normal publications of the respective state publishing systems. The question of reports and their treatment in medical libraries has been considered by Sargeant (1969) who notes that many reports lead directly to journal articles. There is however a growing amount of grey literature in medicine and biology other than the traditional research and development document, as will be seen later.

The breakdown of barriers between scientific disciplines has led to the appearance of more and more medical and biological information being cited in the major grey literature announcement services and the extent to which the topics feature is shown by a comparison of three important publications.

Firstly, *Government reports announcements and index* (GRA & I) published by the National Technical Information Service (NTIS) has three major headings in the NTIS subject category and subcategory structure, namely:

(1) Biomedical technology and human factors engineering, with subcategories in instrumentation, bionics and artificial intelligence, life support systems, protective equipment, prosthetics and mechanical organs, and tissue preservation and storage;

(2) Health planning and health services research, with subcategories in a wide range of health care activities and administration;

(3) Medicine and biology, with subcategories in anatomy, biochemistry, botany, clinical chemistry, clinical medicine, cytology, genetics and molecular biology, dentistry, ecology, electro-physiology, immunology, microbiology, nutrition, occupational therapy, parasitology, pathology, pest control, pharmacology, physiology, psychiatry, public health, radiobiology, stress physiology, surgery, toxicology and zoology.

This extremely comprehensive subject list emphasizes GRA & I's role as a major source of items of grey literature in the medical and biological fields, and indeed the NTIS 100 best seller list 1974–1984 features no less than eight medical and biological titles in the top twenty items. The *Annotated alphabetic list of medical subject headings* (better known as MeSH and compiled by the National Library of Medicine, Bethesda, Maryland) came top with the highest number of copies sold, and the 1988 version was published in July 1987 with the report number PB87–214235/GAR. Other MeSH top sellers are the *Tree structures* and *Permuted medical subject headings*. Medical titles in demand apart from bibliographical aids are exemplified by *Legionnaires: the disease, the bacterium and methodology* (PB297957), and *Morphology of diagnostic stages of intestinal parasites of man* (PB297962).

The second announcement service to be considered, *Scientific and technical aerospace reports* (STAR) published by the National Aeronautics and Space Administration (NASA) uses the following headings:

51 Life sciences (general);

52 Aerospace medicine, including physiological factors, biological effects of radiation, and the effects of weightlessness on man and animals;

53 Behavioural sciences, including psychological factors, individual and group behaviour, crew training and evaluation, and psychiatric research;

54 Man/system technology and life support, including human engineering biotechnology, and space suits and protective clothing;
55 Space biology, including exobiology, planetary biology and extraterrestrial life.

These headings indicate an applications-oriented coverage, and form the basis of a growing specialization in the life sciences, aerospace medicine and biology. The publications in this area have grown to such an extent that NASA finds it appropriate to publish a continuing bibliography, which is issued in the *Special publications* series, e.g. NASA–SP–7011, supplement 288 (N86–32088). The bibliographies give details of studies arising from NASAs programme, as for example *Biomedical results from Skylab* (N77–33780) and *Space gerontology* (N83–16018), as well as items from elsewhere around the world.

The third announcement service under consideration is *Energy research abstracts* (ERA), from the office of Scientific and Technical Information of the United States Department of Energy (DOE). The abstracts on medicine and biology are arranged under three main headings:

55 Biomedical sciences, basic studies, including behavioural biology, biochemistry, cytology, genetics, metabolism, medicine, microbiology, morphology, pathology, physiological systems, and public health;
56 Biomedical sciences, applied studies, including radiation effects, thermal effects, chemicals metabolism and toxicology, and other environmental pollutant effects;
57 Health and safety (no subcategories).

In addition various environmental issues are covered under a further four headings. Many of the items noted in ERA refer to radiation and environmental problems, as for instance *Electron beams in radiation therapy* (13:16170); *Biological indicators of radiation exposure* (13:16196); and *Health effects associated with indoor air pollutants* (13:16016).

The foregoing announcement services are American in origin: in the United Kingdom items in the grey literature concerned with biology and medicine are monitored by the staff at the British Library Document Supply Centre and announced in *British reports translations and theses* (BRTT), using the classification system previously employed by NTIS, that is the COSATI scheme, section 06 of which deals with the biological and medical sciences. The subdivisions cover biochemistry, bio-engineering, biology, bionics and artificial intelligence, clinical medicine, environmental biology, escape, rescue and survival, hygiene and sanitation, industrial medicine, life support systems, hospital equipment, microbiology, personnel (medical), pharmacology, physiology, protective equipment, radiobiology, stress physiology, toxicology, weapon effects, genetics, cytology and molecular biology – in fact a mixture of subjects large and small. Many of the entries under headings such as biochemistry and clinical medicine relate to PhD theses, whilst the heading industrial (i.e. occupational) medicine carries details of translations originating with the Health and Safety Executive, London.

The German publication *Forschungsberichte aus Technik und Naturwissenschaften* covers some medical topics, and the reports which it announces are available from the Central Library for Medicine in Cologne.

Diversity of interests

Grey literature in medicine and biology ranges from the highly specialized (e.g. *Treatment of degenerative diseases of the spine by physiotheraphy*, a translation from the German issued by the Defence Research Information Centre, Orpington, and identified as N87–29108/GAR), to matters of public concern (e.g. *The radiological accident in Goiania*, published by the International Atomic Energy Agency in Vienna in 1988). Many publications are announced and discussed in the general press, especially those from pressure groups, charities and specialized institutions dealing with issues at the uneasy interface between professionals and laymen. Titles representative of this type of grey literature include *Aids* (International Planned Parenthood Federation); *Care of dying children* (National Association of Health Authorities); *Electricity for life? choices for the environment* (Council for the Protection of Rural England); *The nation's health: a strategy for the 1990s* (on preventable causes of death) (King Edward's Hospital Fund for London); and *Crying baby – how to cope* (Great Ormond Street Hospital).

The diversity of such material, the amount of which shows no signs of abating, emphasizes more than ever the need for proper bibliographical control and ease of access. Some materials are indeed adequately covered in standard reference sources, provided the researcher remembers to check them. For instance a great number of biological and medical publications emanate from government departments, and are recorded in Chadwyck-Healey's *British official publications not published by HMSO*, perhaps not the obvious source for details of *Biotechnology in Japan* (Science and Engineering Research Council) or *International response to drug misuse* (Foreign and Commonwealth Office).

Finally, medicine provides an excellent example of the working of the Freedom of Information Act referred to in Chapter 2, namely the ability to access unpublished reports and other information contained in the files of the United States Food and Drug Administration (FDA). The FDA is responsible for ensuring the safety and efficiency of all pharmaceuticals and medical devices introduced into interstate commerce, and in order to discharge this responsibility, reviews approval applications for new products, tracks product performance after approval, inspects manufacturing facilities, and provides guidance for manufacturers in complying with regulations. The information accumulated as a result of these activities is of great public interest, and an organization called FOI Services Inc publishes a source book of FDA Regulatory Documents, and has over 70 000 FDA documents in a collection which began in 1975. FOI uses the Freedom of Information Act route to draw on three main sources for its information:

(1) Center for Drug Evaluation and Research CDER;
(2) Center for Biological Evaluation and Research CBER;
(3) Center for Devices and Radiological Health CDRH.

Agriculture and food

The reluctance of scientists in medicine and biology to use reports literature as a medium for reporting the results of their work is not shared by their colleagues in agriculture and food. For almost two hundred years accounts of research and development and advisory or extension work have been published for the benefit of all kinds of workers in the British primary industry – agriculture. Originally publication was in the journals and papers of the agricultural societies (for example, Bath and West, 1877–; the Royal Agricultural Society, 1840–) and for nearly one hundred years as separate monographs and reports.

Depending on the stage of development in a country, food production is more or less closely linked with agriculture – at subsistence level, virtually one and the same activity; at a highly sophisticated agricultural level they are quite separate. In the literature, apart from food technology in the extreme, there is quite a degree of overlap, and differentiation between the two disciplines will be avoided here.

There is a considerable wealth of grey literature available in agriculture and food, and it has been well represented in the information explosion as progressively more funds and personnel have been employed in research and development. The Publications pour out from a wide variety of organizations: the Food and Agriculture Organisation of the United Nations; the Organisation for Economic Co-operation and Development; governmental and university departments of agriculture; agricultural research bodies; and research stations general in scope or devoted to one or more of the component disciplines of the subject field.

Because the subject content of agriculture is 'broadly' applied biology, the pattern of production and publication of reports and other items of grey literature is world-wide on a zonal or regional basis rather than purely national. Climate, geological conditions, vegetation, pests and diseases overrun national boundaries and contribute to the broad pattern of cultivation and consumption. These topics are so closely concerned with national economics that there is considerable official activity, and publication in all countries is similar. As a result most of the material is recorded in national bibliographies, catalogues of most governments' publications, and to a lesser extent in trade bibliographies.

The whole of the literature of agriculture (conventional and non-conventional) is an area which has been subject to intense and extensive activity. Not only are there excellent guides such as Lilley's *Information sources in agriculture*, noted above; outstanding international announcement services such as the *Bibliography of agriculture*; and specific reference aids like Bush's *Agriculture: a bibliographical guide* (1974); there is also the International Association of Agricultural Librarians and Documentalists (IAALD). The IAALD has looked at the grey literature in agriculture and food in considerable depth, and an overview by Chillag (1982) chronicles the attempts which have been made to measure the extent of the material and come to an agreement on how it should be handled. He quotes Burntrock (working in the early 1970s) as referring to the existence of over 500 secondary services in agriculture, and that on average the same document may be abstracted in seven publications. Chillag says boldly, 'make AGRIS work', and no doubt the members of IAALD will ponder his message.

AGRIS, the International Information System for the Agricultural Sciences and Technology, was established following recommendations at the 1971 conference of the Food and Agriculture Organisation. It covers agricultural literature world-wide, including fisheries, food sciences, forestry and veterinary science. The approximate number of references it contains from 1975 to the present day is a million and a half. AGRIS is the most recently established of the abstracting and indexing services which aim at coverage of all branches of agriculture, and it is also available in a monthly paper version called AGRINDEX. It is the intention of AGRIS to provide for the input of non-conventional material by national and regional contribution centres, hence the exhortation noted above. Currently AGRIS appears to be working satisfactorily for the material originating in the Third World countries and Eastern Europe, but problems have still to be overcome with respect to the output from the United Kingdom and United States, which of course already have their own well established services.

A co-operative venture on a narrower basis is AGREP, a permanent inventory of agricultural research projects in progress in the countries of European community.

Announcement services

Announcement services will be considered in the same sequence as for medicine and biology; firstly there is *Government reports announcements and index* (GRA & I). The NTIS subject category and subcategory structure uses three headings, as follows:

(1) Agriculture and food, with subcategories in agricultural chemistry, agricultural economics, agricultural equipment, facilities and operations; agricultural resource surveys; agronomy, horticulture and plant pathology; animal husbandry and veterinary medicine; fisheries and aquaculture; and food technology;
(2) Environmental pollution and control, with a subcategory on pesticides pollution and control;
(3) Natural resources and earth sciences, with subcategories on forestry and soil sciences.

Typical entries include *Pesticide assessment guidelines* (PB 83153 916) and *Bibliography of the flat grain beetle* (PB88–187893/GAR).

In *Scientific and technical aerospace reports* (STAR), coverage of agriculture and food is, as one would expect, somewhat restricted. The headings used by NASA vary, and it is necessary to look in a number of places, as the following examples show:

(1) Aeronautics (general) for a report on agricultural aviation research (N78–12999);
(2) Earth resources and remote sensing for a report on production estimates for cotton and soy beans (N78–21568);
(3) Meteorology and climatology for a report on a comparative statistical study of long-term agroclimatic conditions affecting the growth of US winter wheat (N81–22658);

(4) Man/system technology and life support for a report on food service and nutrition for the space station (N85–24733).

The Department of Energy's *Energy research abstracts* (ERA) conceals the heading agriculture and food technology as the final subsection of category 55: Biomedical sciences – basic studies. References are scattered throughout the publication, and include many on acid rain. Specific topics range from *Solar energy for a hot air dryer at the Lie farmstead in the county of Flesberg* (13:14866) to *Radioactivity – a layman's guide* issued by the British Food Manufacturing Industries Research Association (13:16947).

In Great Britain, BLDSCs *British reports translations and theses* (BRTT) uses category 02: Agriculture, plant and veterinary sciences to cover agricultural chemistry, agricultural economics, agricultural engineering, agronomy and horticulture, animal husbandry, forestry, veterinary sciences, and fisheries and aquaculture. Food technology appears under Category 06: Biological and medical sciences. Material on agriculture and food is relatively thinly represented in BRTT and typical titles concern pig herd management (88–08–02E–004) and bacterial pathogens in food (88–08–06H–002).

In Germany *Forchungsberichte . . .* announces items of agricultural interest, and the literature supply organizations are the Central Library for Agricultural Sciences at the University of Bonn, and the Veterinary College Library at Hanover.

The major bibliographical services in food and agriculture do include a certain amount of grey literature. For example the International Food Information Service (IFIS) covers, (in addition to many hundreds of journals, books and patents), reports, conference proceedings, specifications and legislation in its three publications:

(1) *Food science and technology abstracts* (FSTA)
(2) *Packing science and technology abstracts* (PSTA)
(3) *Viticulture and enology abstracts* (VITIS/VEA)

IFIS was founded in 1968 and is sponsored by CAB International (UK); the Bundesministerium fur Ernahrung, Landwirtschaft und Forsten (Germany); the Institute of Food Technology (USA); and the Centrum voor Landbouwpublikaties en Landbouw documentaties, PUDOC (Netherlands). Abstracts are edited in the United Kingdom and transmitted to Berlin for computer processing.

The announcement journals of CAB International (formerly the Commonwealth Agricultural Bureaux), collectively known as *CAB Abstracts* and currently consisting of twenty-seven titles, constitute the only major agricultural database with abstracts. *CAB Abstracts* include nearly two million records since 1973, and take in over 14 000 journals, plus books, conference proceedings and reports. Underpinning the whole activity is the *CAB Thesaurus*, which was first published in 1983, with a new edition in 1988. It represents the world's largest English language thesaurus for agricultural and related sciences and is used to index *CAB Abstracts* and also AGRICOLA, the database of the (US) National Agricultural Library. It contains some 56 000 terms.

AGRICOLA (Agricultural On-Line Access) is a family of data files with indexes to world journal and monographic literature, and to US technical reports on agriculture, agricultural economics, food and nutrition, and related topics. AGRICOLA, formerly called CAIN, consists of a number of sub-files, including indexed material from the American Agricultural Economics Documentation Center, the Food and Nutrition Education and Information Materials Center, and several other specialized agencies. The printed version of AGRICOLA is *Bibliography of agriculture*, an announcement service dating back to 1942.

Other sources

As with biology and medicine it is possible to find references to grey literature in a range of sources. Certainly reference should be made to Chadwyck-Healey to identify titles such as *Developments in food marketing in the USA* (Department of Agriculture for Northern Ireland) and *Protected landscapes* (Countryside Commission). Matters of food and agriculture are of great concern to the public at large, and many grey literature documents dealing with topics of the day are noted in the general press. Typical of such items are the *Future of rural communities* (Association of District Councils); *Bovine Somatotropin: a product in search of a market* (i.e. chemical growth boosters) (London Food Commission); *European Community farm policy (National Consumer Council); and Who can afford to live in the countryside?* (from the charity Acre – Action with Communities in Rural England, and the Royal Agricultural College).

Summary

Existing announcement and abstracting arrangements for conventionally published life sciences literature appear to be adequate. Grey literature however does not seem to be so satisfactorily covered. For example, in agriculture, only relatively few grey literature document entries can be found in AGRINDEX, and even these can be attributed to just a few input centres.

Formal notification of life sciences grey literature has a long way to go yet, compared with what has been achieved in energy or aerospace grey literature, both of which have some interlinks with the life sciences. It is true of course that in energy, especially nuclear energy, and in aerospace, non-conventional documents always predominate, but there are lessons which can be learnt. In the case of *Scientific and technical aerospace reports* (STAR), access to documents is gained centrally at the National Aeronautics and Space Administration (NASA). With the International Nuclear Information System (INIS), the input derives from national centres.

There are advantages in processing input to such databases at access points through which very large number of non-conventional literature pass. A lot of fringe material will inevitably be missed by centres which collect material only in narrow subject fields. What seems to be of prime importance is that life sciences databases are not starved of grey literature input just because it is thought that such input requires special expertise. The examples of nuclear energy and aerospace have already been mentioned and this chapter has demonstrated that their respective announcement services have items of great relevance to workers in the life sciences.

The choice lies between the sectorial approach, sending details of grey literature items to the appropriate data bases in the life sciences, or the general approach, sending documents to national centres (the British Library Document Supply Centre in the case of Great Britain) for publication in *British reports theses and translations* (BRTT) and subsequent incorporation in the SIGLE databases. Of course, the solution can be a combination of both approaches.

References

Breen, N. (1988) Environmental information online. *European Environment Review*, **2** (1) 19–28

Bush, E.A.R. (1974) *Agriculture: a bibliographical guide*. London: Macdonald and Janes

Chillag, J.P. (1982) Non-conventional literature in agriculture – an overview. *IAALD Quarterly Bulletin*, **27**, (1) 2–7

Deschamps, J.A. (1986) International and national government information services. Chapter 4 of *Toxic hazard assessment of chemicals*, M.L. Richardson (ed.) London: Royal Society of Chemistry

Gilbert, M. (1988) ECDIN: a European databank on dangerous chemicals. *European Environment Review*, **2** (1) 35–36

Lilley, G.P. (1981) (Ed.) *Information sources in agriculture and food science*. London: Butterworths

Morton, L.T. and Godbolt, S. (1984) (Eds.) *Information sources in the medical sciences*, 3rd edn. London: Butterworths

Sargeant, C.W. (1969) Handling technical reports in the medical library. *Bulletin of the Medical Library Association*, **57**, (1) 41–46

Wood, D.N. (1973) Editor. *Use of earth sciences literature. London: Butterworths*

Wyatt, U.V. (1987) Editor. *Information sources in the life sciences*. London: Butterworths.

CHAPTER NINE

Business and economics

Introduction

At the general level, economics and business studies, although they are classed as part of the social sciences, have much in common with science and technology in the structure of the literature and the way in which it is used. In particular periodical articles have increased in importance and in recent years, working papers, essentially unpublished drafts of potential periodical articles, have achieved a more essential role. The general literature in this area has been covered very thoroughly in two other works in this series, namely *Information sources in economics* (Fletcher, 1984) and *Information sources in management and business* (Vernon, 1984), and consequently the present chapter will confine its attention to the grey literature, which includes the following categories of publications:

(1) Theses;
(2) Research and development reports;
(3) Working papers;
(4) Market research reports;
(5) Reports from banks and stockbrokers.

It is worth pointing out that the problems of disseminating information about, tracing the origin of, and obtaining copies of items of grey literature are relatively acute in the social sciences, partly because the material is of relatively recent growth and partly because so far there are few bibliographical tools.

Theses

The main bibliographical works for theses described in Chapter 5 give comprehensive coverage to economics and business subjects, and it is sufficient here to note the special features of each in its treatment of the social sciences area. The major source is *Dissertation abstracts international* (DAI), section A of which deals with the humanities and social sciences, and subsection 5 specfically covers the social sciences. A closer look at subsection 5 reveals that among other subjects it covers business administration (embracing accounting, banking, management and marketing) and economics (dealing with agriculture, commerce-business, finance, history, labour and theory). DAI is published by University Microfilms International, which also issues *American doctoral dissertations* (ADD) and *Masters abstracts* (MAI), both of which compilations should be consulted for information on business and economics.

Theses originating in Great Britain and Ireland are recorded in the Aslib *Index* described in Chapter 5; the heading for economics has a number of subdivisions. More specifically, it is possible to trace business and economics theses through a number of journals, notably the December issue of *American economic review*; the British publication *Economic journal*; and the March issue of the *Journal of economic history* which carries extended abstracts of recent economic history theses.

Research and development reports

The bulk of government funding for research and development tends to be in the scientific and technical areas, but the major announcement journals, including *Government reports announcements and index* (GRA & I) *Scientific and technical aerospace reports* (STAR) and *Selected Rand abstracts* (RAND) each have sections devoted to business and economics. This is not particularly surprising in the case of GRA & I, which comes under the US Department of Commerce. The relevant section is, under the old COSATI system, field 05: behavioural and social sciences, and under the newer NTIS subject category and subcategory structure: administration and management, and business and economics.

Two related headings are problem-solving information for state and local governments; and urban and regional technology and development. The NTIS 100 best seller list for the period 1974–1984 features a number of titles in the business and economics area, for example *Small business guide to Federal R&D funding opportunities*

(PB83192401); and *Information and steps necessary to form research and development limited partnerships* (PB83131516).

In STAR, heading 83 is devoted to economics and cost analysis, but broader issues can appear under different headings; for example the *Proceedings of the second symposium on space industrialisation* (N85–11011) discuss economic factors and space commercialization and are entered under Astronautics (General).

Reports issued by RAND, an independent, non-profit organization engaged in scientific research and analysis include the *Financial cost of export credit guarantee programs* (R–3491) and *Cost-based reimbursement for nursing home care* (P–7353–RGS).

For reports on business and economics issued in the United Kingdom the prime source of information is the British Library's *British reports translations and theses* (BRTT). The arrangement of the entries follows the COSATI scheme and sections 05A: Management, administration and business studies, and 05D: Economics and economic history give details of documents originating in a wide variety of organizations, including universities, foundations, consultants and charities. The subject matter ranges from the profitability of horse racing (88–08–05D–002) to the determinants of fixed capital investment (88–08–05D–041). Government funded research and development in the business and economic area in the United Kingdom is channelled through a number of agencies, one of the most important of which is the Economic and Social Research Council (ESRC). The ESRC, established by Royal Charter in 1965, supports research and training in the social sciences in British universities, polytechnics and research institutes. Its reporting requirements specify that end-of-award reports, preferably in the form of succinct documents of not more than 5000 words in length, will be deposited in the British Library Document Supply Centre and thus made available for public borrowing. In addition the award holders authorize the Council and BLDSC to disseminate reports by copying, microfilm, microfiche or other means. This procedure represents a definite effort to ensure that the ESRCs research results are clearly directed along the correct routes for grey literature. In addition of course, details are given in the ESRCs own publications, such as *Research supported by the ESRC* and *ESRC newsletter*.

Working papers

The working paper or discussion paper has a special significance in business and economics, and Fletcher (1984) has recounted how one eminent academic described the sequence of publication as 'first a draft paper circulated to a small select group of colleagues upon whose discretion he could rely if the paper was bad; second a revised draft duplicated in sufficient quantities (usually a few hundred) to send to interested individuals and organizations; and finally a manuscript submitted for publication in a learned journal'.

Such a sequence takes time, especially if due allowance is made for colleagues' comment, and the result is a considerable body of information in circulation in a less than well organized manner. (Despite the foregoing remarks, some working papers *are* published – as working papers; see for example *Working for patients*, a set of eight working papers on the future of the National Health Service, published by HMSO in 1989, and dealing in particular with the economics of hospital treatment.)

However, thanks to a commendable initiative by the University of Warwick library, which recognized the importance of economics working papers and the problems which they created for research workers and librarians alike, a collection was begun at Coventry in 1968.

Since then over 70 000 items have been processed, and the current collection consists of about 25 000 items (half in hard copy and half in microfilm). The collection contains materials in economics and management, and from other areas where material is related to economics and management, such as demography, statistics, development studies and law. Papers are received from other universities, polytechnics and colleges, business schools, research institutions, banks and government and other official organizations. The collection is international in scope, with around 600 institutions contributing papers. Over 6500 items are received annually and after four years papers are deposited at the British Library Document Supply Centre, with the exception of the university's own series and some other categories, which are retained permanently.

An author index to the collection is provided by a card catalogue, which also includes the material deposited at BLDSC. The subject index takes the form of a computer print out.

For many years working papers have been listed in bibliographies produced by the library staff and published in conjunction with the Trans-Media Publishing Company. The oldest title is *Economics working papers bibliography* (EWPB), which began in 1973. In 1982 it was joined by the *Management and accounting working papers bibliography* (MAWPB). Both bibliographies have author, series and subject indexes, and they only include papers received at the University within one year of the publication date of the paper. In addition accession lists of papers received at the university are produced each week; these are circulated internally and also published in *Contents of recent economic journals* (HMSO) and *Contents pages in management* (Manchester Business School). Other libraries issuing such accessions lists are Harvard Business Library and the European Institute for Advanced Studies in Management. Further details are available from the Secretary of the Working Papers Collection at the University.

Working papers are not the exclusive domain of the businessman and the economist; they are also much favoured by politicians and officials at the international level, and an instance of their widespread use is to be found at the United Nations, where several important series are made available on subscription. Among them are the *International business, commerce, trade and development papers*, produced by the United Nations Conference on Trade and Development (UNCTAD); *Industrial development papers*, produced by UNIDO (United Nations Industrial Development Organisation), and the mimeographed documents of the United Nations General Assembly and the Security Council. The UNCTAD and UNIDO output includes reports and statistics which are not available anywhere else, whilst the working papers from the General Assembly and the Security Council consist of the provisional records of meetings, preliminary drafts of reports and letters, and draft resolutions.

In Europe the European Parliament issues Research and Documentation Papers, for instance the *Action taken series I–II* 'Achievement of internal market: action taken by the Commission and Council on Parliament's opinions' (1988); and also various *Session documents*. For those who need to exploit collections of European Communities documents, there is an annual index to COM documents (Commission of the European Communities, 1988). COM documents comprise draft legislation (directives and regulations) which pass from the Commission to the Council of Ministers, and reports from the Commission, either to the Council or the European Parliament. Many of these documents are subsequently published either in full or in part, and they are in fact a constituent of the vast and growing output of publications in Europe. Much of it cannot be classed as grey literature because it is made readily available through normal publishing and distribution channels. Nevertheless considerable areas of uncertainty remain, and this is reflected in the establishment of the Association of European Documentation Centres Librarians, one of the aims of which is a programme of training seminars on the European Communities and their publications, directed at the increasing number of people from a variety of backgrounds who are having to come to grips with an undoubtedly complex area.

Commercial publishers too see an opportunity for continuously updated information, especially in view of the added urgency created by the move towards the 'single market', and the part work *Croner's Europe* (1988) is an attempt to meet this need.

Market research reports

As a form of grey literature, market research reports have a number of special characteristics which distinguish them from other publications in the genre. Firstly they are expensive or relatively so; secondly, (like some of the products they deal with) they have a short shelf life; and thirdly they are regularly announced and discussed in the daily press. It is possible to find items with a cover price of £2 500 (*Beyond the 1990s gold rush*: Metals and Minerals Research Services, 1988); items dealing with short-lived products (*Meat and meat products*: Marketing Strategies for Industry, 1988); and items given the headline treatment (*Fast food fad belies the meatless myth*: Guardian account of a Mintel report, 1988).

Market research reports and market surveys are regularly noted in *British reports translations and theses* (BRTT), in section 05D: Economics and economic history, and the British Library (1988b) has published an important compilation called *Market research: a guide to British Library holdings*. Market research reports are excellent sources of ready-worked market information, presenting analyses of data on a market sector, product or service in a structured and authoritative way which can be of immediate use to an enquirer, and in many cases do away with the need for specific desk and field surveys.

The British Library Science Reference and Information Service (SRIS) has a collection of over 2000 reports on a broad range of topics, built around publications issued by a core group of organizations, including:

(1) Economist Intelligence Unit;
(2) Euromonitor;
(3) Infotech;
(4) Jordans;
(5) Key Note Publications;
(6) Market Assessment;
(7) Marketing Strategies for Industry (MSI);
(8) National Economic Development Office.

Other reports are selected on an individual basis.

In addition to the *Guide* mentioned above, there are a number of finding aids to assist the enquirer in the search for a particular item. Probably the most useful is *Marketing surveys index* (MSI), which is regularly updated and also available online; the American *Findex: the directory of market research* (New York, Find/SVP); and *Marketsearch* (London, BOTB/Arlington Management).

Market research reports are of course noted in the many periodicals published in the business and economics area, and two of the most helpful guides are *Reports index* (Dorking, Business Surveys), and its companion publication *Research index*, which together survey a whole range of official and non-official publications and more than one hundred UK newspapers and periodicals.

Reports from banks and stockbrokers

Of all the grey literature documents in the business and economics area, reports based on research carried out by stockbrokers are probably the most transitory, primarily because they are produced for clients with the object of influencing investment decisions. Nevertheless whilst such reports do remain current, they are invaluable sources of information about the financial world and are frequently quoted and commented upon. The BLSRIS receives reports from three major firms, namely Barclays de Zoete Wedd, Alexanders Laing and Cruickshank, and Morgan Stanley. The results of stockbroker research are also available online. Stockbrokers' research features particularly prominently in the run-up to privatization sales; for example the firm of Phillips and Drew prepared a 78–page analysis in October 1988 of steel production around the world, in order to demonstrate the British Steel Corporation's profitability.

Reports from banks on the other hand have a more lasting value, especially those from the world's central banks, which are an important source of up-to-date financial, economic and statistical information on the countries which issue them. The reports comprising the collection at the Joint Library of the International Monetary Fund and the World Bank, Washington, are available on microfiche as *Annual reports of the world's central banks* (Cambridge, Chadwyck-Healey). The year 1984 constitutes the base collection, and updates are available for subsequent years, as is a retrospective collection covering the period 1946–1983. The World Bank itself publishes its own report, and individual banks in all countries publish items which properly belong to the grey literature – an example is the National Westminster Bank's work of reference *Official sources of finance and assistance for industry (1988)*.

Statistical information

Statistical information is a subject which has been reviewed extensively elsewhere, as for example Allott's (1984) contribution to *Printed reference material*; the important point she makes is that whilst 'the majority of libraries still obtain the greater part of their statistics in conventional printed form . . . the situation is changing'. One of the changes is CRONOS, a numerical databank which combines published and unpublished information, some files of which are updated on a daily basis and some files of which are issued as serial publications. CRONOS is produced by Eurostat – the Statistical Office of the European Communities based in Luxemburg, which means that figures are received from all the Member States. The data in CRONOS are official and for the majority of the contributors the statistics come from their national statistical offices. In addition figures are included from the USA, Japan and other countries.

CRONOS contains around one million time series, which have been divided into a number of areas corresponding to CRONOS publications. The areas are:

(1) General statistics;
(2) Economy and finance;
(3) Population and social conditions;
(4) Energy and industry;
(5) Agriculture forestry and fisheries;
(6) Foreign trade;
(7) Services and transport.

Access to CRONOS can be via publications, tapes and diskettes, or on-line through three hosts, WEFA (Paris), DC (Valby, Denmark) and GSI–ECO (Paris). Whether statistical data can be categorized as literature, let alone grey literature, is open to question; what is certain is that CRONOS contains a mix of information available through normal publishing channels and information so new that it is only accessible on-line. The move towards the integration and harmonization of databases will continue, making possible direct comparisons between different countries and facilitating the further development of econometric modelling.

Summary

As noted at the outset, the grey literature of business and economics mirrors to some extent that of science and technology;

the one major absentee is a comprehensive announcement journal covering all types of publication in the manner of say *Government reports announcements and index*. The volume of material is growing and in addition to the categories discussed above, mention must also be made of conferences, meetings and symposia, all of which give rise to papers and proceedings of the kind examined in Chapter 5. Company reports too are a rich and varied source of information, issued primarily for the enlightenment of shareholders and employees, but always available on request from the company secretary's office. Many collections exist, notably the one at the Cranfield Institute of Technology, and that containing the annual reports of the top 200 UK companies maintained by the British Library Science Reference and Information Service. Indeed the reader can do no better than consult the succinct guide to the whole collection (British Library, 1988a).

References

Allott, A.M. (1984) Statistics as a reference source In *Printed reference material*, G.L. Higgens (ed.) 2nd edn, London: Library Association

British Library (1988a) *Business information: a brief guide to the reference resources of the British Library, 2nd edn. London: BL*

British Library (1988b) *Market research: a guide to British Library holdings*, 5th edn. London: BL

Commission of the European Communities (1988) *Index to 1987 COM documents of the Commission of the European Communities*. London: Eurofi

Croner's Europe (1988) London: Croner Publications

Fletcher, J. (1984) (Ed.) *Information sources in economics*, 2nd edn. London: Butterworths

Vernon, K. (1984) (Ed.) *Information sources in management and business*. London: Butterworths

CHAPTER TEN

Education

Introduction

All the perspectives of study discernible in grey literature in education employ their own particular terminology to make a point. The same terms have different meanings much more often than in the language of science and a tendency for the idiosyncratic use of conceptual language reflects the long tradition of individual research in education and allied subjects. In education the tangle of terminology is made denser still by differences between national systems of education and schools of thought. Furthermore educational research, whether of the extended, or snapshot survey kind becomes obsolete through social changes beyond its control rather than through the momentum of its own progress, as in science.

Important studies in education, as was noted in the first edition of this work (Davies and Gwilliam, 1975), are certainly published much less frequently in technical report series than are studies in conventional scientific fields. Even when they are, they do not exhibit the typical features of consistent serial organization and laboratory production, instanced even in work on education's foundation subjects such as applied psychology, psychometrics and so on.

The conventional scientist probably has greater incentive to disseminate his findings and greater opportunity to do so formally.

However the importance of informal contacts for economical but adequate information exchange among educational workers should not obscure the fact that similar kinds of communication are typical of all scientific fields. The place of the unpublished document in educational studies (traditionally somewhat lacking in formal organization) is conditioned by the investigators' concern to control, if not restrict, the personal exchange of information in ways which seem to be the most productive for them, and also the most conducive to an understanding and informed use of the results reported.

There is probably a large proportion of reports in education with small technical content, many dealing with current development, administrative and policy matters, and some attempting to inform non-specialists about teaching and educational research and their mutual relevance. Many reports and other unpublished documents have an essentially local value; many quickly become obsolete. Yet collectively such items become more valuable to historians as social source documents than as scientific and technical reports, if only because education comprehends and reflects so many aspects of social evolution and social need, including the changing face of science and technology.

In concluding this brief look at the background to information and its treatment in education, reference must be made to the book by Dibden and Tomlinson (1981) called *Information sources in education and work*, which adopts a somewhat different approach from other works in the guides to *Information sources* . . . series in that it confines its attention to publications and sources likely to help everyone interested either in starting a career or involved with assisting those who have to make career decisions.

Document sources

The origins of reports and other items of grey literature in education can be traced to official publications at national and local government level issued from the mid-nineteenth century onwards. At about this time, too, educational studies had become autonomous subjects at certain universities. As with other subject areas, much of the most useful grey literature material receives more formal publication at a later stage, but of course a great deal does not. Furthermore although there is a substantial continuity in original sources, many documents derive ad hoc from particular projects, programmes and individual efforts. An examination of the educational scene reveals that the following types of organizations issue various sorts of grey literature:

(1) National public bodies – such as the Department of Education and Science; the Economic and Social Research Council; and the Schools Curriculum Development Council;
(2) Local public bodies – such as local education authorities;

(3) Foundations – the great charitable organizations including Nuffield, Ford, Leverhulme, Calouste, Gulbenkian and Van Leer;

(4) Universities and colleges – especially those establishments with departments of education, psychology and/or sociology, and particularly those institutions possessing separate schools of education;

(5) Regional and international organizations – for example the United Nations Educational Scientific and Cultural Organization, (Unesco);

(6) Associations and professional organizations, of which there are many – for instance the *Education authorities directory and annual* has a section called *Educational associations, societies and other organizations concerned with education*, and in 1988 the list ran to seven hundred and twenty seven different bodies, mostly issuing publications of one sort or another. The names of the organizations ranged from the expected (Headmasters' Conference; Maria Montessori Training Organization; and the Scottish Council for Research in Education) to the not-so-expected (English Mini-Basket Ball Association; Keep Britain Tidy Group; and the Royal Philatelic Society).

Announcement services

The grey literature relating to education, mainly but not exclusively that originating in the United States, is announced in the monthly *Resources in education* (RIE), sponsored by the Educational Resources Information Center (ERIC) of the Office of Educational Research and Improvement (OERI), Washington. ERIC is a USA wide information network for acquiring, selecting, abstracting, indexing, storing, retrieving and disseminating education related documents considered timely and significant. It consists of a co-ordinating staff in Washington, and sixteen clearinghouses located at universities or professional organizations across the United States. The clearinghouses, each responsible for a particular educational area, are an integral part of the ERIC system. Their names are:

(1) Adult, career and vocational education (CE);
(2) Counselling and personnel services (CG);
(3) Educational management (EA);
(4) Elementary and early childhood education (PS);
(5) Handicapped and gifted children (EC);

 (6) Higher education (HE);
 (7) Information resources (IR);
 (8) Junior colleges (JC);
 (9) Languages and linguistics (FL);
(10) Reading and communication skills (CS);
(11) Rural education and small schools (RO);
(12) Science, mathematics and environmental education (SE);
(13) Social studies/social science education (SO);
(14) Teacher education (SP);
(15) Tests, measurement and evaluation (TM);
(16) Urban education (UD).

RIE consists of abstracts (termed résumés) and indexes. The resumes (*Figure 10.1*) provide detailed descriptions of each document, and are numbered sequentially by an accession number beginning with the prefix ED (for ERIC document). The indexes provide access by subject, personal author, institution and publication type. The last index is of particular interest because it characterizes documents by their form or organization, as distinct from their subject matter, and so presents an overview of the components of grey literature in education. The categories are shown in *Figure 10.2*.

The documents cited in RIE except where otherwise noted, are available from the ERIC Document Reproduction Service (EDRS) in both microfiche and paper copy, or in some cases microfiche only. In the United Kingdom ERIC documents can be obtained from the British Library Document Supply Centre.

Most of the documents noted in RIE are channelled via one of the sixteen clearinghouses, but each issue of RIE also carries an open invitation to submit documents considered suitable for announcement, direct to the ERIC Processing and Reference Facility.

As *Figure 10.1* shows, the ED entries constitute very comprehensive records; nevertheless it is interesting to compare them with those in other announcement services such as *Scientific and technical aerospace reports* (STAR), where great pains are taken to ensure that a document's original identifier (report number, series number etc) is not overlooked. In RIE this does not always appear to be the case, for it is possible to find entries relating to documents originating outside the USA, but for which no orginator's reference is cited, simply an ED number. See for example ED 283937 *The role of vocational adult education in promoting the successful employment of women: a British perspective.*

The subject coverage of RIE ranges from academic freedom to youth problems, and the institutions listed in the index include the

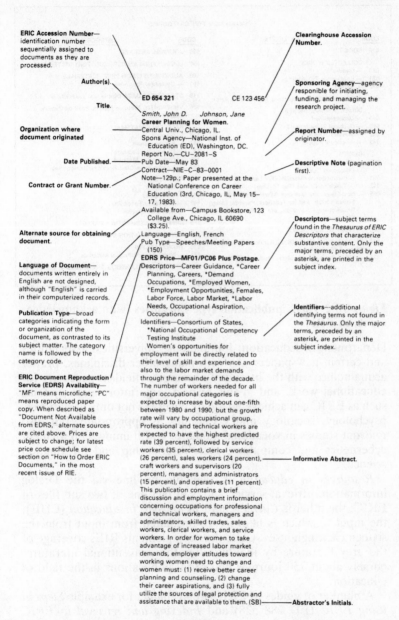

ERIC Accession Number—identification number sequentially assigned to documents as they are processed.

Author(s).

Title.

Organization where document originated

Date Published.

Contract or Grant Number.

Alternate source for obtaining document.

Language of Document—documents written entirely in English are not designed, although "English" is carried in their computerized records.

Publication Type—broad categories indicating the form or organization of the document, as contrasted to its subject matter. The category name is followed by the category code.

ERIC Document Reproduction Service (EDRS) Availability—"MF" means microfiche; "PC" means reproduced paper copy. When described as "Document Not Available from EDRS," alternate sources are cited above. Prices are subject to change; for latest price code schedule see section on "How to Order ERIC Documents," in the most recent issue of RIE.

Clearinghouse Accession Number.

Sponsoring Agency—agency responsible for initiating, funding, and managing the research project.

Report Number—assigned by originator.

Descriptive Note (pagination first).

Descriptors—subject terms found in the *Thesaurus of ERIC Descriptors* that characterize substantive content. Only the major terms, preceded by an asterisk, are printed in the subject index.

Identifiers—additional identifying terms not found in the *Thesaurus*. Only the major terms, preceded by an asterisk, are printed in the subject index.

Informative Abstract.

Abstractor's Initials.

ED 654 321   CE 123 456

Smith, John D.   Johnson, Jane
Career Planning for Women.
Central Univ., Chicago, IL.
Spons Agency—National Inst. of
   Education (ED), Washington, DC.
Report No.—CU–2081–S
Pub Date—May 83
Contract—NIE–C–83–0001
Note—129p.; Paper presented at the
   National Conference on Career
   Education (3rd, Chicago, IL, May 15–
   17, 1983).
Available from—Campus Bookstore, 123
   College Ave., Chicago, IL 60690
   ($3.25).
Language—English, French
Pub Type—Speeches/Meeting Papers
   (150)
EDRS Price—MF01/PC06 Plus Postage.
Descriptors—Career Guidance, *Career
   Planning, Careers, *Demand
   Occupations, *Employed Women,
   *Employment Opportunities, Females,
   Labor Force, Labor Market, *Labor
   Needs, Occupational Aspiration,
   Occupations
Identifiers—Consortium of States,
   *National Occupational Competency
   Testing Institute
   Women's opportunities for
employment will be directly related to
their level of skill and experience and
also to the labor market demands
through the remainder of the decade.
The number of workers needed for all
major occupational categories is
expected to increase by about one-fifth
between 1980 and 1990, but the growth
rate will vary by occupational group.
Professional and technical workers are
expected to have the highest predicted
rate (39 percent), followed by service
workers (35 percent), clerical workers
(26 percent), sales workers (24 percent),
craft workers and supervisors (20
percent), managers and administrators
(15 percent), and operatives (11 percent).
This publication contains a brief
discussion and employment information
concerning occupations for professional
and technical workers, managers and
administrators, skilled trades, sales
workers, clerical workers, and service
workers. In order for women to take
advantage of increased labor market
demands, employer attitudes toward
working women need to change and
women must: (1) receive better career
planning and counseling, (2) change
their career aspirations, and (3) fully
utilize the sources of legal protection and
assistance that are available to them. (SB)

Figure 10.1 **Sample résumé in** *Resources in education* **(with** acknowledgements to ERIC)

PUBLICATION TYPE CATEGORIES

CODE	CATEGORY	CODE	CATEGORY
010	BOOKS	080	JOURNAL ARTICLES
	COLLECTED WORKS	090	LEGAL/LEGISLATIVE/REGULATORY MATERIALS
020	— General	100	AUDIOVISUAL/NON-PRINT MATERIALS
021	— Conference Proceedings	101	— Computer Programs
022	— Serials		
030	CREATIVE WORKS (Literature, Drama, Fine Arts)	110	STATISTICAL DATA (Numerical, Quantitative, etc.)
	DISSERTATIONS/THESES	120	VIEWPOINTS (Opinion Papers, Position Papers, Essays, etc.)
040	— Undetermined		
040	— Doctoral Dissertations		REFERENCE MATERIALS
042	— Masters Theses	130	— General
043	— Practicum Papers	131	— Bibliographies
	GUIDES	132	— Directories/Catalogs
050	— General	133	— Geographic Materials
	— Classroom Use	134	— Vocabularies/Classifications/Dictionaries
051	— Instructional Materials (For Learner)		REPORTS
052	— Teaching Guides (For Teacher)	140	— General
055	— Non-Classroom Use (For Administrative & Support Staff, and for Teachers, Parents, Clergy, Researchers, Counselors, etc. in Non-Classroom Situations)	141	— Descriptive
		142	— Evaluative/Feasibility
		143	— Research/Technical
060	HISTORICAL MATERIALS	150	SPEECHES, CONFERENCE PAPERS
070	INFORMATION ANALYSES (Literature Reviews, State-of-the-Art Papers)	160	TESTS, EVALUATION INSTRUMENTS
		170	TRANSLATIONS
071	— ERIC Information Analysis Products (IAP's)	171	— Multilingual/Bilingual Materials
		999	OTHER/MISCELLANEOUS (Not Classifiable Elsewhere)

Figure 10.2 **ERIC publication type categories** (with acknowledgements to ERIC)

Department of Education, foundations, universities, associations and certain newspapers. As with many other disciplines, proper acquaintance with the literature of marginal fields is necessary for educational work, and in this respect large information systems such as ERIC can achieve adequate coverage not only of statistics, psychology, sociology and economics as appropriate, but also relevant studies in vocational guidance and training, employment cybernetics and computers, plus librarianship and information science.

Resources in education is available on line via the Dialog information retrieval service and is in fact one of two sub-files of ERIC; the other is *Current index to journals in education* (CIJE), the input to which is likewise co-ordinated from input from the sixteen clearinghouses. CIJE thus complements RIEs coverage of the grey literature by looking after the conventional literature, namely about 750 journals and serial publications in the field of education.

A number of guides to ERIC are available: for example *Steps in using ERIC* (ED 288 528) and *Indexing and retrieval in ERIC* (ED 279 346). The future status of ERIC has been the subject of hearings before the House of Representatives, and papers for public comment on a redesign of the ERIC systems were issued in 1987 (ED 278429).

The sheer size and resources of ERIC dwarf efforts by other agencies to tackle the question of grey literature in education, but the initiative taken by the British Library Document Supply Centre in seeking to co-ordinate items arising from British sources must not be overlooked. The announcement journal *British reports translations and theses* (BRTT) lists items relating to education and allied topics under some of the headings which form part of 05: Humanities, psychology and social sciences, especially the following:

05B Documentation information science and librarianship;
05K Linguistics;
05P Education and training;
05Q Psychology;
05R Sociology.

Topics covered range from the education of pupils of Chinese origin in Leeds (88–08–05P–007) to a report on satellites and education (88–08–05P–021).

The announcement services familiar to workers in science and technology also carry items relating to educational matters. *Government reports announcements and index* (GRA & I) has a place for education under the category Behaviour and society. It also reports documents with an educational content under other headings, notably Library and information sciences, and frequently quotes ED numbers, invariably pointing out that the documents in question have to be obtained from ERIC.

Scientific and technical aerospace reports (STAR) places education under category 80: Social sciences (general) and very occasionally items appear under this heading – see for example *The impact of science on society* (N85–24994). With *Energy research abstracts* (ERA) it is necessary to look hard for material on education: the relevant headings are 09: Education and public relations under section 32: Energy conservation, consumption and utilization; and 03: Information handling, under section 99: General and miscellaneous.

Selected RAND abstracts, a quarterly guide to the publications of the RAND Corporation (RAND derived from research and development) includes items on education, as for example *Effective teacher selection: from recruitment to retention* (R–3462).

The general picture which emerges is that apart from BRTT most announcement services carry very little on education, and are content to leave the garnering of grey literature to ERIC. The approach is emphasized by a further activity, for in addition to collecting the literature of education for announcement in RIE and CIJE, the ERIC clearinghouses analyze and synthesize the literature in a number of different formats designed to compress the vast amount of information available and to meet the different needs of ERIC users. The formats include research reviews, state-of-the-art studies, interpretive studies on topics of high current interest, research briefs, annotated bibliographies, and various compilations. Items range from a three-page digest of the plain English movement (ED 284 273) to a 73–page report on computer-based education in the social studies (ED 284 825).

Some authorities feel that ERIC does not offer an altogether adequate coverage of British research; the reader can however turn to the compilation prepared by the National Foundation for Educational Research called the *Register of educational research in the United Kingdom* (NFER and Nelson, Windsor). The *Register* dates from 1973, and the volume covering the years 1983–1986 was published in September 1987.

As was mentioned earlier in this chapter, a great deal of information on education is published by governments and international organizations; much of it does not belong to the grey literature because it is announced and made available through for example Her Majesty's Stationery Office (HMSO) or the United States Government Printing Office (USGPO) and their respective catalogues. Similarly the publications on educational and cultural policy issued by the European Communities can be identified from the appropriate publications catalogues. If however publication takes place outside such main channels, specialist services need to be checked, and in the case of Great Britain a major source is Chadwyck-Healey's *British official publications not published by HMSO*, which lists titles such as *Public Libraries and adult independent learners* (Council for Educational Technology) and *Career breaks for women engineers and technicians* (Engineering Industry Training Board).

On broader issues the monthly *International labour documentation* published by the International Labour Organization carries details of monographs as well as journal literature, and includes in its subject coverage vocational training and social development.

Also in the context of Europe, reference must be made to EUDISED, the formative stages of which were described by Davies and Gwilliam (1975). EUDISED (European documentation and information system for education) is a project of the Council for Cultural Co-operation of the Council of Europe for a decentralized computer-based education documentation and information network. At the beginning the principal concern was the recording of on-going research, and in particular, coverage of ERIC reports listings. European report sources were to be covered economically by project entries in standard format, organized by a special multilingual thesaurus giving details of the literature associated with the projects. Later it was proposed to enlarge EUDISED beyond the coverage of grey literature by adding details of articles from educational periodicals and monographs. It is now possible to access EUDISED as an on-line database through the host ESA/IRS, but with just 7000 references on file for the period 1975 to the present day (according to the statement in the ESA/IRS fact sheet), it looks as though the early hopes for the project have not yet been fulfilled. Indeed in an appraisal paper by Davies (1980), it is noted that problems exist, namely 'difficulties in getting national co-operation, the necessity of unravelling relationships with other international organizations in a concrete and constructive way with a minimum of overlaps and unnecessary repetitions and competitive ventures, the paucity of staff and the inability to attract appropriate funds for essential activities'. When first conceived the approach to EUDISED was seen as similar to but not identical with ERIC, and the heart of the scheme was a measure to tackle the not inconsiderable language problems, namely the *Multilingual thesaurus for information processing in the field of education*, the first edition of which was published under the aegis of the Council of Europe in English, French and German versions in 1973, in a Spanish version in 1975 and in a Dutch version in 1977. A new edition came out in 1984 (Council of Europe), with the addition of a further four languages – Italian, Danish, Greek and Portuguese. A revised edition is scheduled for the end of 1989.

The EUDISED system itself, which is also available as a printed version called EUDISED *R & D bulletin* and published by K.G. Saur, Munich, has recently been the subject of an independent research evaluation conducted by Johan van Halm and Associates. The evaluation was based on interviews and on questionnaires sent to EUDISED *R & D bulletin* subscribers and to users of the on-line version. A five-point action plan concludes that consideration should be given to ending the EUDISED activities if resources for their efficient development cannot be provided.

In the meantime the EUDISED *R & D bulletin* appears four times a year – each issue contains some 250 reports on ongoing or recently completed projects in the field of educational research and development. Project reports are given in English French or German, and they are indexed with descriptions taken from the EUDISED *Thesaurus*. Grouping within the *Bulletin* is according to the terminographs (French terminogrammes) of the *Thesaurus*. National agencies from twenty-two countries are involved, and those representing the United Kingdom are the National Foundation for Educational Research in England and Wales, and the Northern Ireland Council for Educational Research.

Public involvement

Education, unlike many of the other topics discussed when reviewing grey literature, is something every reader has experienced at first hand. It is no accident therefore that much of the unconventional, outside the book trade type of literature in education is at the grass roots level – because we have all been through it, therefore we all may have strong views as pupils, parents, possibly as school governors. The daily press finds sufficient interest amongst its readers to warrant regular features on education, and the *Times* considers it worthwhile to publish a couple of educational supplements. There is an ever growing pool of information, comment and opinion. Some publications will be recorded in one of the major databases, abstracted and indexed for subsequent retrieval: a lot will fall into oblivion, ignored because they are transient or trivial. What is collected will depend on a combination of chance and recognition. No one can formulate a strict policy on what to look out for, but librarians and documentalists can influence the scope and coverage of grey literature originating in Britain by encouraging issuing organizations to send copies to the British Library Document Supply Centre for possible inclusion in BRTT.

The sort of titles which are appearing emphasize the wide range of interested parties – for example the Chartered Institute of Public Finance and Accountancy advocates the evaluation of teachers in its report *Performance indicators for schools*; the Equal Opportunities Commission has discovered and written about sex discrimination in West Glamorgan schools; the International Freedom Foundation advocates encouraging children to join voluntary defence training corps in its report *Education for defeat*; the Institute of Economic Affairs want the government to introduce education tax credits; and the *Survey of comics and magazines for children and young people* does not recommend Enid Blyton's *Famous five adventure magazine* because of sexual stereotyping – the list can be extended indefinitely, and no surveys or studies of publications on education should be considered complete without at least taking such items into account. Ideally the grey literature database is the place to look.

References

Council of Europe (1984) *EUDISED multilingual thesaurus for information processing in the field of education*. Berlin: Morton Publishers

Davies, J. (1980) EUDISED – image and reality: a crisis of identity. *Education libraries bulletin*, **23** (3) 1–15

Davies, J. and Gwilliam A.B. (1975) Technical reports in education In *Use of reports literature*, edited by C.P. Auger London: Butterworths

Dibden, K. and Tomlinson, J. (1981) *Information sources in education and work*. London: Butterworths

CHAPTER ELEVEN

Energy

Introduction

Public complacency about energy and the supply of fuel received a sharp jolt as a result of the 1973–74 oil crisis, when the western world realized that the days of 'cheap' fuel were over. Governments around the world turned their attention to energy technology and how it could be used to conserve and better utilize existing resources. Attention was also turned to alternative renewable sources such as wave power and solar energy, and these activities resulted in the establishment of national and international agencies, research and development programmes, and a greater concern for fossil fuel reserves.

The United Kingdom was in a fortunate position in that it was able to establish itself as having the largest energy resource of any country of the European Community, and since 1980 has been self-sufficient in energy in net terms, thanks to the continued growth of off-shore oil production. An increase in technical literature has reflected this surge in activity and will continue to do so, for the picture is still changing as the price of a barrel of oil now begins to move steadily downwards. Such developments have been described in succinct terms by J. Brookes (1985) in her chapter 'Energy technology' in *Information sources in engineering*; by Graham and King (1985) in their chapter 'Nuclear power engineering' in the same work; and on a much broader scale in *Information sources in energy technology* (Anthony, 1988). Reports literature and more lately grey literature has always featured prominently in the energy scene, in the main because of its origins in the nuclear industry. It is necessary to retrace these origins in order to understand the present-day arrangements for handling information and publications dealing with energy in all its aspects.

Nuclear energy

Atomic energy programmes consist of nuclear science and technology defined as follows: nuclear science is the study of the production properties and phenomena of atomic nuclei, subatomic particles, gamma rays and nuclear x-rays; nuclear technology is the application of nuclear science to other sciences and engineering, and conversely the application of other sciences and engineering to the problems of nuclear science. The grey literature discussed here is mainly unclassified reports literature, that is reports without any security, commercial or other restrictions. Reports, as has been noted elsewhere in this volume, are generally identifiable by their alpha-numeric code, but as a rule this may not be sufficient to distinguish a report from a conference paper, from a patent, from a translation, or in some cases from a journal article. For this reason the boundaries of reports literature, including that in the atomic energy field, will vary acording to individual interpretation, and have of course led to the very concept of grey literature.

It is true to say that nuclear science and its literature are unique in at least some respects among the various fields of science. Apart from some fundamental physics aspects, nuclear science was born virtually overnight early in 1939 with the publication of letters to the editor in *Physical review* (1939). During the summer of that year the implications of nuclear fission were spelt out by Einstein to President Roosevelt, and the reasons why atomic research during the war years had to be kept so closely guarded a secret are all too well understood today. The initial work was carried out mainly by scientists from the USA and the UK and its dominions, and the activity was known as the Manhattan Project (Groves, 1962): it was directed towards the production of an atomic bomb. Once the potential of this weapon was understood, it was not surprising that nations worked to continue safeguarding their nuclear knowledge and making documentation of it available only in a closely controlled manner. The dissemination of this information was arranged therefore by means of report literature, a medium particularly suitable for the purpose.

Gradually controls were transferred from military to civilian agencies, and the US Atomic Energy Act of 1946 established the United States Atomic Energy Commission. Its Technical Information Service was given the task of 'dissemination of scientific and technical information as called for by the Act'.

Scientists from the United Kingdom, Canada, Australia, France and other countries, having returned home from their wartime work on the Manhattan Project, became leaders of atomic energy research in their own countries. By 1954 it was realized that the world could face extinction by a nuclear holocaust, and attention was turned more to the exploitation of atomic energy for peaceful purposes. Under the auspices of the United Nations, conferences on the peaceful use of atomic energy were held, and in 1957 the International Atomic Energy Agency was created, with headquarters in Vienna. Today there are 113 member countries participating in the work of the IAEA, whose principal missions are to assist countries in the peaceful uses of the atom in agriculture, health, energy and other fields, and to apply safeguards on nuclear materials and facilities provided through its programmes or upon request from Member States, so as to verify that they are not used for military purposes. In the field of documentation this mission is carried out through the International Nuclear Information System (INIS), of which more later.

United States Atomic Energy Commission

The development and growth of the United States Atomic Energy Commission (USAEC) has been described by a number of authorities, notably Hewlett and co-workers (1962), (1969), Smyth (1945), Fermi (1954), and Crewe and Katz (1969). Its story came to an end in 1974 when the Commission was abolished and its functions were transferred to the US Nuclear Regulatory Commission (NRC) (Murray, 1988).

The reports literature, or any literature for that matter, of atomic energy has its beginnings in 1942. From that date the University of Chicago Metallurgical Laboratory provided a dissemination and central indexing system for the documents arising from the Manhattan Project. During the war the dissemination of this reports literature was very closely guarded indeed. Then the Atomic Energy Act of 1946 established the USAEC. The Army and the Office of Scientific Research and Development (OSRD) relinquished control of the Manhattan Project and the USAEC was left with the task of declassification and release of documents accumulated during the wartime research. The Manhattan District Declassification Center was created and the first reports for public use released (the MDDC and AECD series).

From its inception the USAECs research and development efforts were conducted almost entirely under contract with USAEC-owned but privately operated laboratories, with universities and non-profit institutions, or with commercial enterprises. The contractors were required to report periodically on the results or progress of their work. In basic research much of this reporting took the form of articles in scholarly journals, but the results of sponsored research were more frequently given in reports submitted directly to the USAEC.

Most of such reports, as well as the articles were announced and abstracted in *Nuclear science abstracts*, published semi-monthly. The staff of the USAECs Technical Information Center was charged with the task of carrying out section 141 of the Atomic Energy Act of 1954, which states: 'The dissemination of scientific and technical information relating to atomic energy should be permitted and encouraged so as to provide the free interchange of ideas and criticism which is essential to progress and public understanding, and to enlarge the fund of technical information'. Section 141 also opened wide the door for international co-operation.

In the mid–1950s the USAEC designated over one hundred centres within and outside the USA as depository libraries. Until 1963 the depositories automatically received either hard copy or microcard versions of USAEC sponsored R & D reports announced in *Nuclear science abstracts*. In 1963 the dissemination of reports was changed to the more convenient microfiche.

For various reasons, by the early 1970s the USAEC depository system was gradually phased out. Instead the National Technical Information Service in the USA, the IAEA in Vienna, and the British Library became the main sources of USAEC and other reports, although some of the depositories continued to maintain their collections by purchase.

Energy research abstracts

The most important continuing bibliography covering the energy field is *Energy research abstracts* (ERA). ERA provides abstracting and indexing coverage of all scientific and technical reports, journal articles, conference papers and proceedings, books, patents, theses and monographs (in other words a mixture of conventional and grey literature) originated by the US Department of Energy, its laboratories, energy centres and contractors. ERA also covers other energy information prepared in report form only by federal and state government organizations, foreign governments, and domestic and foreign universities and research organizations. The Department of Energy draws special attention to the fact that ERA coverage of non-report literature is limited to that generated by the Department's own activities.

ERA is compehensive in scope, encompassing the DOEs research, development, demonstration and technological programmes resulting from its charter for broad coverage of energy sources, supplies, safety, environmental impacts and regulation.

ERA is the successor to *Nuclear science abstracts*, which began publication in 1946 as *Abstracts of declassified documents*. It changed its name to NSA in 1948 when its scope was widened to include material other than USAEC reports.

Today ERA is produced as part of the Department of Energy's Scientific and Technical Information Program (STIP), which is carried out at many levels within the Department and by its contractor organizations. The Office of Scientific and Technical Information (OSTI) at Oak Ridge, Tennessee, provides the necessary direction for STIP and serves as DOEs national centre for scientific and technical information management and dissemination. Both DOE originated information and world-wide literature regarding advances in subjects of interest to DOE research workers are collected, processed and disseminated through an energy information system maintained by OSTI. The major data bases in this system are available within the United States through commerical on-line systems and to those outside the United States through formal government exchange agreements.

Typical entries in ERA are shown in *Figure 11.1*. Five indexes are provided for approaching the contents of each issue of ERA, namely:

(1) Corporate author index;
(2) Personal author index;
(3) Subject index;
(4) Contract number index;
(5) Report number index.

The corporate author index lists many well-known establishments around the world, including the Atomic Energy Establishment, Winfrith; the Brookhaven National Laboratory; the Centre d'Etudes Nucleaires; the Rutherford Appleton Laboratory; and the Stanford Linear Accelerator Center. The subject contents are arranged according to a scheme which provides for 40 first-level and 289 second-level subject categories; full details of the scheme's scope definitions and limitations are contained in the report DOE/TIC–4584–R6.

The report number index lists items from all the major reports series. By far the largest series is CONF–, which provides details of conference papers and conference proceedings. A typical issue of ERA will carry details of around 3500 items.

ABSTRACTS IN *ENERGY RESEARCH ABSTRACTS*

The principal elements of abstract entries for a typical research and development report and a typical technical journal article are illustrated below.

Report number   Date of publication   Corporate author

Author(s)   Title   Availability

5785 (BMI/ONWI–522) Thermal property and density measurements of samples taken from drilling cores from potential geologic media. Lagedrost. J.F.: Capps, W. (Fiber Materials, Inc., Biddeford, ME (USA). Dec. 1983. 179p. NTIS, PC A06/MF A01; 1; Order Number DE84004926. GPO Dep.

Legibility Code

Density, steady-state conductivity, enthalpy, specific heat, heat capacity, thermal diffusivity and linear thermal expansion were measured on 59 materials from core drill samples of several geologic media, including rock salt, basalt, and other associated rocks from 7 potential sites for nuclear waste isolation. The measurements were conducted from or near to room temperature up to 500°C, or to lower temperatures if limited by specimen cracking or fracturing. Ample documentation establishes the reliability of the property measurement methods and the accuracy of the results.

Title   Journal citation   Date of publication   Author(s)

15701 Equilibrium and power balance constraints on a quasistatic Ohmically heated field-reversed configuration (FRC). McKenna, K.F.; Rej, D.J.; Tuszewski, M. (Los Alamos National Lab., NM (USA)). *Nuclear Fusion*; **23**: No. 10, 1319–1325 (Oct 1983).

Zero-dimensional power balance calculations are performed for a quasi-static, purely Ohmically heated field-reversed configuration. Without compression, the constraint imposed by radical pressure balance limits the power input. Estimates of the energy loss from impurity line radiation as well as from classical and anomalous transport are given. Effects of cold puff gas injection are also investigated.

Abstract

Figure 11.1 **Typical entries in** *Energy research abstracts* (with acknowledgements to DOE)

In addition to *Energy research abstracts* the Department of Energy also publishes an announcement service limited to non-technological or quasi–technological articles or reports having significant reference value; the service is called *Energy abstracts for policy analysis* (EAPA), and places its emphasis on policy, legislatory, and regulatory aspects; social, economic and environmental impacts; regional and sectional analysis; and institutional factors. The contrast is made with what the Department terms the 'hard' scientific and technical literature in *Energy research abstracts*, and the coverage includes items from the USGPO, NTIS and the Department of Energy itself.

Other announcement services

Whilst ERA is the principal American announcement service on energy matters, other publications also deal with the subject. For example *Government reports announcements and index* (GRA & I) has two main categories:

(1) Energy;
(2) Nuclear science and technology.

It uses a number of subcategories as well, and inevitably there is some duplication, as for instance DOE/TIC–4583–R2/GAR *Energy data base: guide to abstracting and indexing* noted in GRA & I 1986, item 632967.

Scientific and technical aerospace reports (STAR) also caters for documents concerned with energy, under the main heading 44: Energy production and conversion, which includes specific energy conversion systems, e.g. fuel cells; global sources of energy; geophysical conversion; and windpower. In addition the reader is referred to 07: Aircraft propulsion and power; 20: Spacecraft propulsion and power; and 28: Propellants and fuels. NASA works with the Department of Energy (e.g. N83–19231 on large horizontal-axis wind turbines) and also sponsors work of its own (e.g. N78–13527: Solar cell high efficiency and radiation damage).

International nuclear information system

Whilst *Nuclear science abstracts* has broadened into *Energy research abstracts*, the *International nuclear information system* (INIS), operated by the International Atomic Energy Agency

(IAEA) in Vienna has continued to concern itself entirely with the
literature in nuclear science and technology, namely:

General physics	Radioisotope effects
High energy physics	Applied life sciences
Neutron and nuclear physics	Health, radiation protection and
Chemistry	environment
Materials	Radiology and nuclear medicine
Earth sciences	Isotopes and radiation sources
All effects and various aspects	Isotopes and radiation application
of external radiation	Engineering
Fission reactors (general)	Nuclear documentation
Specific fission reactor types	Safeguards and inspection
and their associated plants	Mathematical methods and codes
Instrumentation	Computer codes
Waste management	Miscellaneous (general relevant
Economics and Sociology	documents)
Nuclear law	

INIS is a co-operative decentralized information system set up
by IAEA following the approval of the Board of Governors in
1969. Its purpose is to construct a database identifying publications
relating to nuclear science and its peaceful applications. Member
states and co-operating international organizations scan the
scientific and technical literature published within their boundaries
or by them, select from it those items which fall within the subject
scope noted above, and process the data according to agreed
standards and rules. Document descriptions, abstracts and in
certain cases full texts are then submitted to the IAEA headquarters,
where all the data are merged and the INIS output products
compiled.

The following services are issued:

(1) INIS magnetic tape service;
(2) INIS *Atomindex*, a semi-monthly abstracting journal;
(3) INIS non-conventional literature on microfiche.

Each issue of INIS *Atomindex* consists of a main entry section and
a number of indexes (personal author, corporate entry, subject,
conference and report, standard and patent number). The main
entries are arranged by subject categories, and within each
category, report literature is followed by journals and books.
Cumulative indexes are produced half-yearly and annually; there
is also a ten-year cumulation of report, standard and patent
number indexes covering the period 1977–1986.

Readers of *Atomindex* are expected to obtain items of conventional literature cited (books and journals) through the normal distribution channels of the book trade. However in the case of grey literature (which INIS terms non-conventional literature, including scientific and technical reports, patent documents, pre-conference papers, and non-commercially published theses) the INIS Clearinghouse, a unit within the INIS Secretariat supplies on request and for a given fee, microfiche copies of most of the non-conventional literature announced by INIS. The Clearinghouse databank, called CLIN, covers all non-conventional literature added to the system since it began, and serves as a catalogue of documents whether available on microfiche or not.

Full details of the services available from Vienna can be found in the document *Presenting INIS* (report GEN/PVB/13).

British sources of energy information

Until 1945 responsibility for atomic energy matters in the United Kingdom rested with the Department of Scientific and Industrial Research (DSIR). In that year responsibility was transferred to the Ministry of Supply. The Atomic Energy Research Establishment was set up at Harwell; then the Production Group was established at Risley in 1946; and in 1947 the Weapons Group was created.

Initially the establishments set up under the Atomic Energy Act of 1946 were fully engaged in defence commitments, but gradually the emphasis moved to the peaceful uses of atomic energy. 1954 saw the passing of the Atomic Energy Authority Act, by which piece of legislation the United Kingdom Atomic Energy Authority was created as a public board. The subsequent development of UKAEA in its task of carrying out the research and development necessary to ensure that nuclear power is economic, safe and environmentally acceptable; the creation of British Nuclear Fuels plc to provide nuclear fuel services covering the whole fuel cycle; and the proposal, following a review by the Secretary of State for Energy, to introduce legislation to enable the UKAEA to operate as a trading fund, are outside the scope of this work.

However, as far as information goes, the UKAEA, in accordance with the provisions of the Atomic Energy Acts, makes publicly available the results of its work, except when they need to be protected on the grounds of national or commercial security.

Access to abstracts of UKAEA documents can be gained through *Energy research abstracts* and INIS *Atomindex*. In addition UKAEA publishes its own monthly list of publications, wherein it identifies documents as:

(1) those deposited at the British Library Document Supply Centre;
(2) those on sale through HMSO;
(3) those which have appeared in the literature.

The items sent to BLDSC are recorded in *British reports translations and theses* (BRTT) under section 10: Energy and power. In particular the following sub-headings are used:

(1) 10E fission fuels;
(2) 10O Nuclear power plants;
(3) 10P Nuclear reactor technology.

The broader aspects of energy within the United Kingdom are co-ordinated by the Department of Energy, which has responsibility for policies for all forms of energy including its efficient use and the development of new sources; and the government's relations with the nationalized energy industries and the UKAEA.

The Department is an important source of information on all aspects of all forms of energy, and publishes a catalogue *Publications in print*, supplemented by annual lists. The items listed are typical of documents appearing in the grey literature and include R & D reports sponsored by the Energy Technology Supply Unit (ETSU); *Offshore technology reports; Energy papers*; and major reports arising from the Department's programme on Severn tidal power (STP).

As with UKAEA, the Department's publications are also listed under the appropriate heading in BRTT.

Finally, mention must be made of an important guide (Chester, 1986) to selected literature and sources of information on nuclear energy and the nuclear industry held by the British Library, including that on open access at the Science Reference and Information Service in London, and that available for loan from the Document Supply Centre, Boston Spa. Now in its second edition, the guide covers abstracts, bibliographies, periodicals, patents, business literature and directories.

Public concern

As with other major topics covered in this book (medicine, education), the question of energy is one which touches every individual in the land, and as a result a new class of grey literature is emerging which originates not with government departments agencies and laboratories, but with pressure groups and institutions quite outside the energy industry. Thus the daily press regularly carries details of expressions of concern over issues of the day. A few titles will serve to indicate the general pattern: *The greenhouse effect: issues for policy makers* (Royal Institute of International Affairs); *Earthing electricity* (Centre for Economic and Environmental Development); *Magnox: the reckoning* (Friends of the Earth); *Radiation monitoring around CEGB nuclear power stations* (Institute of Physics); and so on. What permanent value such documents have and how they should be stored and indexed is still a matter for conjecture, but clearly they cannot be ignored in the short term, and may have reference value in the future.

References

Anthony, L.J. (1988) (Ed.) *Information sources in energy technology*. London: Butterworths

Brookes, J.A. (1985) Energy technology. In *Information sources in engineering*, L.J. Anthony (Ed.) 2nd edn. London: Butterworths

Chester, K. (1986) *Nuclear energy and the nuclear industry*, 2nd edn. Boston Spa: British Library

Crewe, A. and Katz, J.J. (1969) *Nuclear research*. New York: Dover

Fermi, L. (1954) *Atoms in the family – my life with Enrico Fermi*. University of Chicago Press, Chicago

Graham, F.A. (1985) Nuclear power engineering. In *Information sources in engineering*, L.J. Anthony (Ed.) 2nd edn. London: Butterworths

Groves, L.R. (1962) *Now it can be told: the story of the Manhattan Project*. New York

Hewlett, R.G. and Anderson, O.E. (1962) *The new world 1939/1946* (Volume 1 of *History of the US Atomic Energy Commission*) Pennsylvania State University Press

Hewlett, R.G. and Duncan, F. (1969) *Atomic Shield 1947/1952* (Volume 2 of *History of the US Atomic Energy Commission*) Pennsylvania State University Press

Murray, R.L. (1988) *Nuclear energy*, 3rd edn. Oxford: Pergamon, – *see especially* Chapter 23: Laws, regulations and organizations

Physical Review (1939) **55**, 416–418, 511–512, 797–800; **56**, 284–286, 426–450

Smyth, H.D. (1945) *Atomic energy for military purposes*. New York: Princeton University Press

CHAPTER TWELVE

Science and technology

Introduction

The literature of science and technology is large by any standard, and the proportion of it represented by grey literature, especially reports, is quite significant. Indeed impressive though the coverage of the conventional literature is through long-established and all-embracing announcement services such as *Chemical abstracts*, the INSPEC series, and *Engineering index*, no survey of the literature of a particular branch of science or technology can be considered complete unless full attention is paid to information contained in the grey categories.

The reasons why abstracting and indexing services covering science and technology have prospered and expanded over the years is because they are vital for the success of research and development programmes which are in turn the driving force of science-based technologies and ultimately affect and influence the progress of whole sectors of industry. Keys to the literature of science and technology have long been recognized as a vital component in both the academic and industrial fields of endeavour, and consequently sufficient money has been forthcoming to enable services to be organized and sold on a self-sustaining basis.

At first when reports and other types of grey literature appeared on the scene, their treatment was somewhat haphazard but as the realization of their value 'as sources of information complementary to the conventional literature sources' grew, efforts were made to co-ordinate the accessioning and announcement of such documents. The result has been a measure of control equal to or even better than that of the published literature, and an attempt will be made here to indicate some of the more important guides.

The field of science and technology is an ill-defined one embracing a great number of disciplines, and noticeable, at least in terms of reports literature, for the important presence of aerospace and energy, subjects which are dealt with separately in Chapters 7 and 11.

The life sciences too account for a great deal of the output of the conventional publishing services, but less so in the case of grey literature; they are considered in Chapter 8.

The actual effectiveness of the use of grey literature in the industrial sector is not easy to gauge. It is one thing for an agency or a government department to specify and promote research and development in a particular project or addressed to a particular problem: terms of reference are agreed, a sum of expenditure is named, and the work is executed and reported upon. It is quite another matter to judge how well companies or indeed sectors of industry adopt the same piece of research and development work for their own purposes. An addition to knowledge it may be but whether it is further exploitable is quite another matter. This problem has of course greatly exercised bodies such as the National Aeronautics and Space Administration, which has made great attempts to stimulate technology transfer across a whole spectrum of industries.

In surveying science and technology to consider its grey literature aspects, it is important to draw attention to a couple of fairly recent relevant titles in the present series, namely *Information sources in physics* (Shaw, 1988), and *Information sources in engineering* (Anthony, 1985). It is also important to mention a work of a comprehensive but somewhat different nature in the field, namely *Information sources in science and technology* (Parker and Turley, 1986), which is a practical guide to traditional and on-line use, and constitutes a reference work aimed at both the organizers and users of scientific and technical information.

Finally, since so much of the grey literature in this area is of American origin, readers will find invaluable the work by Aluri and Robinson (1983) *A guide to US government scientific and technical resources*, an encyclopaedic compilation covering among other topics technical reports, scientific translations, indexes and abstracts, databases and information analysis centres.

National Technical Information Service

In the United States the organization which dominates the grey literature scene, especially the reports scene, is the National

Technical Information Service (NTIS), an agency of the US Department of Commerce, the central source for the public sale of US government-sponsored research, development and engineering reports, and for the sales of foreign technical reports and other analyses prepared by national and local government agencies and their contractors or grantees. In order to understand the present day importance and influence of NTIS it is worth recounting how it all began.

The public law that serves as the 'charter' for NTIS essentially provides the agency with a mandate to disseminate scientific, technical and engineering information to the American public. Details of the relevant Executive Orders and Public Laws are given in Appendix A of the work by McClure and co-workers (1986). The key dates in the history of NTIS can be summarized as:

(1) 1945–1946 Office of the Publication Board (Established by Executive Order 9568:1945);
(2) 1946–1965 Office of Technical Services (OTS) (Established by Executive Order 9809:1946);
(3) 1965–1969 Clearinghouse for Federal Scientific and Technical Information (CFSTI) (Agency placed under the National Bureau of Standards);
(4) 1970– National Technical Information Service (NTIS). (CFSTI abolished and its functions transferred to NTIS).

Today in terms of size the NTIS activities are enormous – the collection is approaching two million titles, several hundred thousand of which contain foreign technology or foreign marketing information. All the titles are available for sale, mostly as copies from microform masters, but about 80 000 titles in more frequent demand are held as shelf stock. Annually NTIS supplies its customers with more than six million documents and microforms – some 24 000 items daily.

Some of the major services based on the NTIS collection are as follows:

(1) NTIS *Abstract newsletter*, published on a weekly basis in 26 subject categories which summarize most unclassified federally funded research as it is completed and made available to the public;
(2) On-line access to the NTIS *Bibliographic database* available through DIALOG Information services and other hosts; the database covers the period from 1964 and contains approximately 1.5 million records;
(3) Published searches, paper-bound versions of online searches conducted on the NTIS *Bibliographic database* in around 3000 topical subject areas;

(4) *Selected research in microfiche* (SRIM), a facility whereby clients can receive the full text of reports in microfiche distributed according to a previously determined user profile;

(5) Center for the Utilisation of Federal Technology (CUFT), an organization created by the Stevenson-Wydler Technology Innovation Act with the aim of playing a catalytic role in increasing private sector use of technologies created by the Federal government. The Center encourages Federal agencies to identify technical reports with potential commercial applications, and then publicizes them by various direct means, including *Tech notes*, a monthly compilation of fact sheets (e.g. PB88–925900/GAR).

Other services offered by NTIS, and more information on those noted above, can be found in the *General catalog of information services* issued from time to time by NTIS.

Government reports announcements and index

The principal vehicle for the notification by NTIS of the reports and other grey literature added to its collection is *Government reports announcements and index* (GRA & I), which is published twice a month and lists more than 2000 bibliographic citations in each issue. It is completed by the publication of an annual index. Like NTIS itself, GRA & I has undergone several identity changes, as can be seen from *Figure 12.1*.

GRA & I is a comprehensive publication covering an extremely wide range of topics. Until December 1986 GRA & I used the NTIS subject category and subcategory structure endorsed in 1964 by the Committee on Scientific and Technical Information (COSATI) of the Federal Council for Science and Technology, described in AD–612 200; COSATI corporate author headings are described in PB198 275.

From January 1987 onwards, GRA & I began to use the NTIS subject category and subcategory structure which can be used for on-line searching, described in NTIS *Bibliographic database guide*, PR253. The COSATI list had 22 numbered categories; the PR253 list has 38 categories (not numbered). For the NTIS subject classification (past and present) see PB270575.

Each issue of GRA & I is divided into:

(1) Reports announcements;
(2) Keyword index;
(3) Personal author index;
(4) Corporate author index;
(5) Contract/grant number index;

Dates	Agency	Title	Frequency
1946–1949	OTS[1]	*Bibliography of Scientific and Industrial Reports*	Weekly
1949–1954	OTS	*Bibliography of Technical Reports*	Monthly
1955–1961	OTS	*US Government Research Reports*	Monthly
1961–1964	OTS	*US Government Research Reports*	2 per month
1965–1969	CFSTI[2]	*US Government Research and Development Reports* (with separate index)	2 per month
1970–1971	NTIS[3]	*US Government Research and Development Reports* (with separate index)	2 per month
1971–1975	NTIS	*Government Reports Announcements (GRA), Indexed by Government Reports Index (GRI)*	2 per month
1975 to date	NTIS	*Government Reports Announcements and Index (GRA & I)*	2 per month

[1] OTS – Office of Technical Services
[2] CFSTI – Clearinghouse for Federal Scientific and Technical Information
[3] NTIS – National Technical Information Service

Figure 12.1 **GRA & I antecedents**

(6) NTIS order/report number index;
(7) Price codes.

Inside the back cover are to be found details of foreign NTIS co-operating organizations.

Each item in each issue of GRA & I is allocated an abstract number, and the documents themselves are usually identified by one of the following report series codes:

(1) AD–A, representing unlimited, unclassified documents originating with the Department of Defense (AD originally meant ASTIA *document*, and ASTIA stood for the Armed Services Technical Information Agency);
(2) DE, representing documents cited in *Energy research abstracts* published by the Department of Energy;

(3) ED, representing ERIC documents originating with the Education Resources Information Center, and not in fact available from NTIS;

(4) N89–, representing NASA documents;

(5) PB89–, representing documents processed by NTIS itself (PB originally meant Publication Board, the ultimate forerunner of NTIS).

Other identifiers are met with, for example TIB, representing items of German origin processed originally by the Technische Informationsbibliothek, Hanover.

The richness and diversity of the grey literature covered so thoroughly in GRA & I can be assessed in two ways. Firstly there is the vast range of document types which are regularly cited, of which the following list is just a sample:

Annual reports	Journal papers
Bibliographies	Leaflets
Briefing reports	Masters' theses
Bulletins	Memorandum reports
Conference papers	Patent applications
Contract reports	Progress reports
Data files	Quarterly reports
Doctoral theses	Research reports
Draft reports	Summary reports
Final reports	Translations
Granted patents	USGPO publications
Internal reports	Working papers

Secondly there is the sheer spread of the NTIS subject category and subcategory structure, such that the A to Z coverage of a typical issue of GRA & I can read:

Aids	Nose bleeds
Beehives	Ocean waves
Cataloguing	Permanent magnets
Dental plaque	Quality control
Explosives	Research management
Flight paths	Sea urchins
Genes	Thermonuclear reactors
Hiroshima	Urethanes
Interviewing	Viruses
Jamming	Water pollution
Kaolin clay	X-ray analysis
Laundry products	Yttrium
Manned space flight	Zooplankton

There is unfortunately a penalty to be paid for such a wide coverage, and that is the significant amount of overlap among the major bibliographic sources for grey literature. For example a study by Copeland (1981) found that *Government reports announcements and index* covered 94.3 per cent of technical reports that appeared in STAR and 78.8 per cent of those that appeared in ERA. Conversely it was conluded that about 41 per cent and 21 per cent of report entries in GRA & I appeared in STAR and ERA respectively. GRA & I also duplicates items originating with the Education Resources Information Centre (ERIC), and items appearing in the USGPO *Monthly catalog*. Whilst at first sight such duplication may appear undesirable, some authorities hold the view that the repetition of entries provides an improved access to the technical reports in question – especially if the user of the announcement services is aware of the extent of the duplication.

RAND

Among the many United States organizations which issue reports and other items of unpublished literature on any scale, the Rand Corporation is noteworthy in that is publishes its own announcement journal called *Selected RAND abstracts*. RAND (derived from Research and Development) is an independent nonprofit organization engaged in scientific research and analysis. It conducts studies in the public interest supported by the US Government, by local and state governments, its own funds derived from earned fees, and by private foundations. The work involves most of the major disciplines in the physical, social and biological sciences, with emphasis on their application to problems of policy and planning in domestic and foreign affairs. The output of RAND titles is not large – typically 250 to 300 items a year – and this enables *Selected RAND abstracts* (SRA), published four times a year, to be issued on a cumulative basis, with issue no 4 (December) containing the complete collection for the year. Several government agencies distribute RAND publications, including NTIS, NASA and ERIC. The subject coverage is wide, as can be judged from the list of bibliographies containing abstracts of RAND publications selected from the current volume of SRA and from earlier RAND research (*Figure 12.2*).

Information analysis centers

In the United States a number of organizations usually referred to as Information Analysis Centers (IACs) have been set up in order

Africa	Management: Personnel, Organizational Theory, and Administration
Arms Control	
Asia	
California	Mathematical Programming: Theory and Applications
China	
Civil Defense	Middle East
Civil Justice	Military Manpower
Communication Satellites	Military Strategy and Tactics
Communication Systems	NATO
Computing at RAND	New York City
Cost Analysis	Nuclear Research
Criminality, Justice, and Public Safety	Operations Research Methods
	Policy Sciences
Decisionmaking	Population
Delphi and Long-Range Forecasting	Privacy in the Computer Age
Education	Program Budgeting
Energy	R&D and Systems Acquisition
Environmental Issues	Regulatory Issues
Europe	SIMSCRIPT and Its Applications
Expert Systems	Smoking, Alcoholism, and Drug Abuse
Game Theory	Space Technology and Planning
Gaming	Statistics
Health Care Costs and Coverage	Systems Analysis: Methods, Techniques, and Theory
Health-Related Research	
Housing	Television and Communications Policy
Human Resources	
International Terrorism	Transportation
International Trade	Urban Problems
Latin America	U.S.-Foreign Relations
Linguistics	USSR
Logistics: Inventory, Spare Parts, Maintenance	Water Resources
	Weather and Climate Studies

Figure 12.2 **Scope of the coverage in *Selected RAND abstracts* (with** acknowledgements to the RAND Corporation)

to accomplish the analysis and communication of highly specialized technologies.

The now defunct Panel on Information Analysis Centers of COSATI has defined an IAC as 'a formally structured organizational unit specifically (but not exclusively) established for the purpose of acquiring, selecting, storing, retracing, evaluating, analyzing and synthesizing a body of information and/or data in a clearly defined specialized field or pertaining to a specific mission with the intent of compiling, digesting, repackaging or otherwise organizing and presenting pertinent information and/or data in a form most authoritative, timely and useful to a society of peers and management'.

The IACs are normally staffed by scientists who analyze and evaluate published literature, grey literature, and data of all kinds to produce critical data compilations, state-of-the-art reviews, bibliographies, current awareness services and newsletters. Many of these products are in fact grey literature and available through NTIS; others make their appearance in conventional journals or through normal publishing channels. The Department of Defense sponsors a number of IACs and in a recent overview of forty years experience with them noted that whilst they are not without their problems, they continue to have a significant role in the transfer of scientific and technical information (Rothschild, 1987). Details of the activities of a selected list of IACs appear in Chapter 12 of the work by Aluri and Robinson (1983), as does information on other IAC reference works.

Defence

One of the major sources of grey literature in the United States is the Defense Technical Information Center (DTIC), described by Lahr (1981) as 'but one part of a complex bureaucracy which comprises the DOD technical information program'. The DTIC, which serves the Department of Defense and its contractors as well as other US governmental agencies and their contractors, can trace its origins to an effort established in London (UK) in 1945 which was designed to process confiscated German documents. In 1951 the documentation centres involved – the Navy Research Section of the Library of Congress and the Central Air Documents Office at Dayton, Ohio, operated by the US Air Force – were consolidated to become the Armed Services Technical Information Agency (ASTIA). In 1963 the name was changed from ASTIA to the Defense Documentation Center (DDC), and a further name change at a later date resulted in the current title, DTIC.

Whereas NTIS deals only with unclassified and unlimited distribution reports, DTIC handles classified and limited distribution reports as well. As a result, access to DTIC services is restricted and it requires prior registration and certification of a 'need to know'. The main services and products of DTIC include:

(1) a large collection of reports in subject areas such as aeronautics and missile technology as well as in the basic sciences;
(2) an announcement service *Technical abstract bulletin* (TAB), with indexes;
(3) A series of data bases which can be accessed by the *Defense RDT & E online system* (DROLS).

Although the DTIC does not serve the general public directly, the scientific and technical community does have access to reports and other documents having no security or distribution restrictions. Such items are sent to NTIS where they are announced through GRA & I. This policy has some interesting consequences and is not without its critics. Thus the Director of the US Defence Technology Security Administration is quoted as saying 'During the Carter administration a system was set up within the US government to get rid of excess classified documents. The US Defense Department was sending everything but classified information to NTIS. But there was an automatic seven year 'clock' on declassification. When a document was seven years old it was automatically declassified. The documents were about everything, tank warfare, missile fuels, electro-optics, advanced computer databases, radar absorption. They were all dumped at NTIS when they were automatically declassified. So the Soviets became the best customers of NTIS, (Fisk, 1988).

Presentations on the work of DTIC are made at the *Annual users' conferences*, the proceedings of which are notified through GRA & I (see for example AD–A 166 250/1/GAR).

A vast amount of information on defence matters in the United States and around the world is provided by the Jane's Information Group, a consortium founded in 1988 to bring together the resources of Jane's Publishing Inc, DMS Inc and Interavia S.A. Files are compiled from open, unclassified sources, but some, such as File 988, DMS *Market intelligence reports* (MIR) have restricted access.

The UK and European scene

At one time the (UK) Department of Technology operated a Technology Reports Centre (TRC) which processed and made available exploitable and unpublished research and development reports arising from United Kingdom government programmes and those of overseas governments. The TRC held the majority of non-classified technical reports produced in government research establishments, and in some cases the reports series went back to 1940. The Centre's holdings were particularly strong in electronics, aeronautics, materials technology, and mechanical, electrical and industrial engineering. The Centre also received the openly available publications provided by NTIS and NASA, plus many documents from US and overseas organizations under exchange arrangements.

TRC had a host of means for publicizing the contents of its collection, the most important of which was the twice-monthly

R & D abstracts, arranged according to the COSATI scheme. Documents cited could be obtained as paper copies or microfiche, and comprehensive indexes were produced on a half-yearly basis.

However, as the collection at TRC developed it became apparent that there was an increasing overlap between the reports activities of the Centre and those of the British Library at Boston Spa (BLDSC). A decision was taken that TRC should withdraw from reports handling and reliance should be placed on the British Library, with its far greater resources and economies of scale. Consequently TRCs stock of T-numbered reports for the period 1972–1981 was transferred to Boston Spa, where they were made available to registered users. However technical reports and other documents with a defence classification continued to be handled by the Defence Research Information Centre (DRIC), described below. The decision to end TRCs reports handling activities meant the demise of *R & D abstracts*, the last issue of which appeared on 15 December 1981. The file still retains a reference value, not least because the abstracts were very full and informative.

It should also be noted that whereas *R & D abstracts* represented the comprehensive approach to reports announcements in Britain, a more selective method was that embodied in an alerting service originally called TECHLINK. The idea was to exploit specific titles, particularly those considered to have useful industrial applications. A series of illustrated one-sheet digests was prepared by the TRC staff, with a follow-up facility available to subscribers who considered any of the ideas summarized of sufficient interest to warrant further enquiry. However, with the closure of TRC there was no obvious government home for TECHLINK. Thus it was renamed Tech Alert as a service providing summaries of new ideas for industry in engineering and technology and is now operated by Microinfo Limited. Original source material, as previously, is available on request. Microinfo is a firm of publishers and distributors of specialized information, and the organization to which NTIS refers British and Irish enquiries and orders for its products and publications. Microinfo publishes a regular sales bulletin called *NFM (News from Microinfo)* which highlights certain topical PB reports and other documents and services. Requests for microform copies of NTIS publications identified from *Government reports announcements and index* can however still be submitted direct to the British Library Document Supply Centre.

The main announcement journal for unrestricted British reports and other publications is now *British reports translations and theses* (BRTT), a bibliography published monthly (with indexes which cumulate annually) by the British Library Document Supply

Centre. BRTT seeks to cover all material falling within the grey literature category, and lists British report literature and translations produced by British government organizations, industry, universities and learned institutions, and most doctoral theses accepted at British universities and polytechnics during and after 1970. It also covers reports and unpublished translations from the Republic of Ireland and selected British official publications that are not published by HMSO.

BRTT uses the COSATI scheme to arrange its entries, and it is noticeable how section 05: Humanities, psychology and social sciences has expanded in recent years as the concept of grey literature has been widened outside the relatively narrow fields of science and technology. Unlike *R & D abstracts*, BRTT entries carry a basic amount of bibliographic information, no summaries and an augmented title where appropriate, so that the item JET: evolution, status and prospects (86–12–20J–001) is augmented as European tokamak development. A typical BRTT entry is shown in *Figure 12.3*.

An analysis of the December 1986 issue reveals the following items by category:

Theses	*Reports*	*Translations*	*Others*[1]	*Total*
387	277	126	492	1282
30.2%	21.6%	9.8%	38.4%	100%

1. 'Others' includes – background papers, census digests, discussion papers, local plans, occasional papers, working papers.

The issue also contains the following microfiche:

(1) Keyterm index Jan–Dec;
(2) Author index Jan–Dec;
(3) Report number index Jan–Dec.

On the question of publishing reports and other documents, it should not be overlooked that some organizations still prefer to make their own individual arrangements. Workers in specific subject areas will no doubt be aware of the services available in their particular fields, but transport provides two excellent examples, one in the United Kingdom and one in the United States. The (UK) Transport and Road Research Laboratory (TRRL) issues a regular digest of TRRL reports, giving two-page summaries (with illustrations and tables) of items available for sale. The annual output is small, but the system is direct and up to date.

The Transport and Road Research Laboratory is the research establishment of the Department of Transport, which together with the Department of the Environment (which also has a

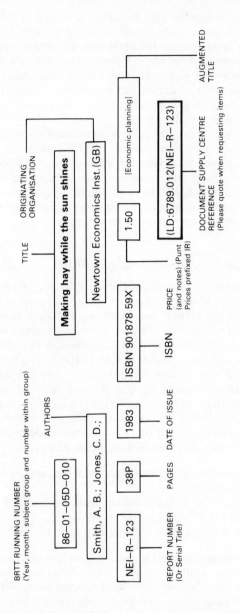

Figure 12.3 **Arrangements of entries in *British reports, theses and translations*** (with acknowledgements to the British Library Document Supply Centre)

research establishment, the Building Research Establishment) issues an annual list of publications detailing documents emanating from the two departments. Information is given about publications issued or sponsored by the Departments, and about research projects sponsored. Generally speaking all the publications are available to the general public, but the sources of supply vary from HMSO to specific government offices – a good mixture of conventional and grey literature.

In the United States, the Transportation Research Board, a body whose purpose is to stimulate research concerning the nature and performance of transportation system, issues a whole series of reports on planning and administration, designs, materials and construction, maintenance and equipment, operations and safety, soils geology and foundations, and transportation law, all of which are listed in the Board's *Publications catalog*.

Defence Research Information Centre (DRIC)

Whereas with the closure of the Technical Reports Centre and the ending of *R & D abstracts*, much of British industry was left to its own devices in the identification and acquisition of grey literature in science and technology, in the case of companies engaged on defence contracts, practical help continued in the shape of the Defence Research Information Centre (DRIC). The basic function of DRIC is to disseminate scientific and technical information, especially that contained in unpublished reports, to the UK defence community, defined as Ministry of Defence branches and establishments, armed service units, military colleges, and UK firms and organizations working on government defence contracts. DRIC adds to its collection of documents in three ways:

(1) Direct from the originator (MOD branches and establishments, and contractors);
(2) from overseas defence information centres, in particular from the USA, Canada and Australia;
(3) as a result of requests for specific reports.

Current holdings comprise one million documents, many of which are held as multiple copies and may range from two pages to several hundred in length. The stock is increasing at the rate of about 10 000 documents per year, split approximately equally between UK and overseas material. DRIC supplies 14 000 reports annually in response to requests, and in addition distributes a further 40 000 copies in accordance with instructions on the incoming documents.

DRIC has an announcement service, *Defence research abstracts* (DRA), which is published in two main editions:

(1) MOD edition – confidential (and available to MOD staff only);
(2) Contractors' edition – restricted (and available to MOD contractors).

Reports from DRICs collection are available to MOD staff and defence contractors on loan, or for retention where the number of copies permits. Non-MOD requesters can approach DRIC, but are advised to quote some *prima facie* evidence of need-to-know (for example a government contract number).

Since it is official policy that wherever possible the results of UK defence research and development shall be applied in the civil field, efforts are made to ensure that any MOD reports received by DRIC which can be given an unlimited distribution are passed to the British Library Document Supply Centre, to be added to its grey literature collection; to the publishers Chadwyck-Healey for inclusion in their *Catalogue of British official publications not published by HMSO*; and to UK copyright libraries.

DRIC issues a helpful pamphlet *A guide to services*, available on request from its offices in Glasgow. Two other publications of interest are the booklet *Selling to the MOD*, aimed at helping small firms and component suppliers to compete for Ministry of Defence business; and the *MOD contracts bulletin*, published by Longmans in co-operation with MOD to open up the UK defence market by giving full details of all invitations to tender.

Euro abstracts

An announcement service published by the Commission of the European Communities is *Euro abstracts*; section 1 covers Euratom and EEC research and development and demonstration projects described in scientific and technical publications and patents. Section 2 deals with research conducted by the European Coal and Steel Community (ECSC).

Euro abstracts carries summaries of four categories of publications:

(1) Reports in EUR series;
(2) Articles published in journals;
(3) Papers presented at conferences;
(4) Patents.

Instructions on how to order, and from where, are given in nine languages.

Summary

As noted at the beginning of this book, reports literature grew out of the need for scientists and engineers to communicate the results of research and development quickly, cheaply and efficiently, with security controls if necessary. Technical reports now constitute a considerable part of the grey literature but paradoxically their role, impact and importance are still the subject of conflicting opinions. An attempt has been made by McClure (1988) to summarize current views on:

(1) Use of the technical report literature by the R & D community;
(2) Manner in which the technical report assists the R & D community;
(3) Assessing the value of technical reports.

He suggests further lines of investigation to try and get an increased knowledge of the subject. In the meantime the fact remains that in science and technology, reports form an important medium for furthering the processes of research and innovation, the transfer of scientific and technical information, and the achievement of technology transfer. Moreover the fact also remains that the grey literature of science and technology depends heavily on US agencies and announcement services for its effective dissemination and bibliographic control. European efforts, notable through *British reports translations and theses* and Germany's *Forschungsberichte* are making increasingly important contributions, but users still rely heavily on *Government report announcements and index, Scientific and technical aerospace reports*, and *Energy research abstracts*.

References

Aluri, R. and Robinson, J.S. (1983) *A guide to US government scientific and technical resources*. Littleton, Colorado: Libraries Unlimited
Anthony, L.J. (1985) (Ed.) *Information sources in engineering* 2nd edn. London: Butterworths
Copeland, S. (1981) Three technical report printed indexes: a comparative study. *Science and technology Libraries*, **1**, Summer, 41–53
Fisk, R. (1988) *The Times* 8 October
Lahr, T.F. (1981) *A study on decreased technical reporting in the Department of Defense*. AD–180 238
McClure, C.R. (1988) The Federal technical report literature: research needs and issues. *Government information quarterly*, **5**, u1, 27–44
McClure, C.R., Hernon, P., and Purcell, G.R. (1986) *Linking the US National Technical Information Service with academic and public libraries* Norwood, N.J.: Ablex Publishing Corporation

Parker, C.C. and Turley, R.V. (1986) *Information sources in science and technology: a practical guide to traditional and on-line use.* 2nd edn. London: Butterworths

Rothschild, M.C. (1987) *Information Analysis Centers in the Department of Defense* AD–A184002/4/GAR.

Shaw, D.F. (1988) (Ed.) *Information sources in physics.* London: Butterworths

APPENDIX A

Keys to report series codes

Several publications are available which give help in identifying and verifying report series codes. Some works attempt to cover as many codes as possible whilst others concentrate on specific subject areas. All are invaluable in helping to recognize and deal with reports codes which are increasingly quoted in the open literature without any form of explanation. The characteristics of the major works of this nature are noted below.

Probably the best known is the *Report series codes dictionary* (Aronson, 1986), the purpose of which is to identify, and provide an association for, most of the codes which have been applied to reports. The *Dictionary* was originally the result of a spare-time effort on the part of members of the (US) Special Libraries Association, with the first edition appearing in 1962, and the second in 1973. The current third edition (1986) is the outcome of work by the Commerce, Energy, NASA, Defense Information Cataloging Committee, Washington, and now runs to 647 pages. It constitutes a guide to over 20 000 report series codes used in the dissemination of scientific, technical and industrial information by nearly 10 000 corporate authors. The *Dictionary* is arranged in two parts, the first alphabetically by report acronym or other identifier, and the second by organization name, with the corresponding report series codes.

A work which complements the *Dictionary* is the *Directory of engineering document sources*, first published in 1973 under the editorship of Simonton; the latest update appeared in 1985, with a slight change in the title from *Directory* to *Dictionary*. Taken together the two dictionaries represent the best general approach to resolving reports codes and other problems of identification, and constitute essential reference works in any grey literature collection.

In specific subject areas too it is often possible to refer to specialized guides. For example the Office of Scientific and Technical Information of the United States Department of Energy has issued *Report number codes*, a compilation in which Part I is alphabetical by report codes followed by issuing organizations. Part II lists the issuing organizations followed by the assigned report code or codes.

References

Aronson, E.J. (1986) *Report series codes dictionary*, 3rd edn. Gale Research, Detroit

Dictionary of engineering document sources (1985). Global Engineering Documents, Santa Ana, California

Nelson, R.N. (1985) *Report number codes*. US Department of Energy DE85005834/GAR

APPENDIX B

Trade literature

Trade literature possesses many of the attributes of grey literature, notably pamphlet format, controlled distribution and varying standards of bibliographical control. Whether it should be treated as part of the grey literature, in that it is collected and records of it added to databases, is not an easy question to answer. Certainly it cannot be ignored, but to admit it as a category alongside other categories such as reports, translations, theses, and local and central government documents is a problem not readily resolved. On the one hand there are enormous quantities of trade literature published each year, making any attempt to organize it time-consuming and expensive: on the other hand certain types of trade literature eventually become of great interest to historians, especially those working in the fields of technology and social studies. The consignment of trade literature to an appendix in the present work reflects the uncertainty.

The purpose of trade literature is to promote sales and to provide information about products and services, functions which some authorities feel are better carried out in separate documents but which in reality are often combined. Thus a typical piece of trade literature will consist of a brochure firstly extolling the virtues of the product or service in terms of its advantages and economics, and secondly stating in the form of a specification the product or service's salient facts and figures. Usually trade literature in library collections is filed for its information content and the sales aspect has no permanent value. The amount of information provided in trade literature will depend firstly on the profile of the likely customer, and secondly on the complexity and nature of the product or service. Typically it will include performance data, operating characteristics and materials properties. In certain areas, notably electronics, trade literature is highly structured and extends to reports on the application of discrete devices and circuits. Usually each commercial organization produces its own trade literature in accordance with how it sees the needs of the moment, and variety rather than uniformity is often a deliberate policy.

Trade literature is often regarded as an extremely valuable, up-to-date and detailed source of technical information, and many methods have been advocated to achieve its full utilization. In some cases it is treated as part of a library's total collection and it is accorded an appropriate amount of cataloguing indexing and storage facilities. The British Library Science Reference and Information Service (SRIS) has a collection of current trade literature from about 20 000 mainly British companies covering a broad range of manufacturing industries, comprising:

(1) Product literature (catalogues, data sheets, price lists, and so on);
(2) House journals;
(3) Company annual reports;
(4) Stockbrokers' reports;
(5) Exhibition catalogues.

These types of literature are interfiled and arranged alphabetically by company name. In addition, SRIS maintains a collection consisting of trade literature published between 1830 and 1940 from about 7500 companies.

Two characteristics of trade literature which undermine any formal attempt to organize it into collections similar to other library material, such as for example reports, are (a) that it is easily and readily available to bona fide enquirers direct from companies and organizations only too willing to provide details of their products and services in the hope of making a sale, and (b) that it is regarded by recipients as expendable simply because it has no price tag attached. Notable exceptions occur, again in electronics, where trade literature is available on a subscription basis. Even when pieces of trade literature do bear a price on the cover, this fact is often ignored by the manufacturer's representative, who supplies the documents free of charge at his discretion.

Not surprisingly therefore many information departments prefer to concentrate on collecting and maintaining buyers' guides and trade directories which point the way to specific products and processes, and leave the actual accumulation and subsequent disposal of the trade literature itself to individuals or technical departments and offices. SRIS has an extensive collection of trade directories, and displays in the library itself copies of a *Subject index to directories*. For the guidance of outside users, SRIS also publishes a holdings list of directories (Science Reference and Information Service, 1986a).

Thus the traditional starting point in obtaining trade literature is the consultation of a trades directory, which may simply be a list of names and addresses, combined with a classification in which companies are listed under product and service headings. A good and long-established example is *Kelly's business directory*, a comprehensive guide to British industry and commerce, whose 102nd edition appeared in 1989.

A different approach is that adopted by Kompass, which emphasizes the geographical spread of manufacturing and supplying organizations, and also employs a series of tables carrying symbols to indicate in detail a particular company's products or services. Other trade directories concentrate on specific subject areas; generally they are arranged in the same way as the comprehensive directories, listing companies alphabetically and under product and service headings. Some publications, particularly those of American origin, actually include extracts from manufacturers' catalogues as an integral part of the directory. Thus the *Diesel and gas turbine worldwide catalog* allocates several pages to each company listed and is therefore able to present a sample of trade literature from each concern.

Periodicals too are an important source of trade literature, or of information on how to obtain it. Many periodicals are available on a restricted distribution basis and are sent free to scientists, engineers, buyers and others considered able to influence the design and/or purchase of components and materials. Periodicals which regularly contain trade directory information are detailed in *Trade directory information in journals* (Science and Information Service, 1986b).

By far the most common way of making available trade literature, as might be expected, is for companies to issue single documents such as brochures and bulletins, which may sometimes be collected together and reissued as manufacturers' catalogues particularly in loose-leaf binders. In many such instances the contact is between a manufacturer's or supplier's representative and the sales prospect, and no involvement with a library or information service is considered necessary. In fact the system can be wasteful in terms of time and storage space since a representative may call on several members of the same organization and establish several different files of literature at various locations. To overcome this problem a number of concerns have examined the idea of offering a one-stop service in the provision of trade literature. One of the most enduring is the service provided by Technical Indexes Limited.

This company maintains a whole series of full-text information files, and each can be provided in microform (on fiche or film) on an annual rental basis including comprehensive indexing and regular updating. Rapid access to required documents is via printed indexes called *Product data books*. The Technical Indexes collection of trade literature includes:

(1) British catalogue files;
(2) Canadian catalogue file;
(3) US Vendor catalogue services.

Technical Indexes also maintains extensive collections of industrial, military and international standards.

In some industrial sectors, notably building and construction, trade literature has for a long time been subject to a high degree of control and organization. One example is the *Barbour compendium of building products*, which lists over 5400 manufacturers supplying a full range of products across 3000 product categories, and carries over 1100 pages of product illustrations in full colour.

A further example is the series of *D.A.T.A. books* published by DATA Incorporated, and designed to report comprehensively on what is presently being produced throughout the world in various sectors of the electronics industry, as for instance interface devices. The enquirer is able to search under:

(1) Electrical and mechanical requirements;
(2) Type number;
(3) Generic or functionally equivalent device number.

Attempts to standardize the content and especially the format of trade literature have been made from time to time. The British Standards Institution issued BS1311 in 1955 on sizes of manufacturers' trade and technical literature (including recommendations for contents of catalogues). Its influence has been minimal, to judge from the multiplicity of paper sizes in any trade literature file, and the standard has in fact been withdrawn, to be replaced by BS4940 which makes recommendations for the presentation of technical information about products and services in the construction industry.

Good presentation of technical literature depends on a number of key features:

(1) Good technical description;
(2) Good quality illustrations;
(3) Integrated layout and design;
(4) Adequate document identification.

Whatever methods (if any) of standardization for trade literature are adopted, and whatever methods are advocated for its handling and exploitation, the prime purpose of trade literature will always remain the furtherance of business rather than the preservation of information or the furtherance of knowledge. Consequently issuing organizations of all kinds will continue to be flexible and choose whichever policy proves the most beneficial to the business.

Further reading on the treatment of trade literature is to be found in the contribution by Wall (1986) to *Information sources in engineering*, and in the SRIS publication by Dunning (1985).

References

Dunning, P.M. (1985) *Trade literature in British libraries*. London: British Library
Science Reference and Information Service (1986a) *Directories held by the Science Reference and Information Service*, 2nd edn. London: British Library
Science Reference and Information Service. *Trade directory information in journals*, 6th edn. London: British Library
Wall, R.A. *Product information: trade catalogues. In Information sources in engineering*, (Ed.) L.J. Anthony. Chapter 7. London: Butterworths

APPENDIX C

Organizations mentioned in the text

Advisory Group for Aerospace Research & Development, AGARD, 7 rue Ancelle, 92200 Neuilly sur Seine, France

American National Standards Institute, ANSI, 1430 Broadway, New York NY 10018, USA

American Society of Mechanical Engineers, ASME, United Engineering Center, 345 E 47 Street, New York NY 10017, USA

Aslib, the Association for Information Management, Information House, 26–27 Boswell Street, London WC1N 3JZ, UK

Association of European Documentation Centres (EDC) Librarians, c/o George Edwards Library, University of Surrey, Guildford GU2 5XH, UK

Association for Information and Image Management, AIIM, 1100 Wayne Avenue, Suite 1100, Silver Spring, MD 20910, USA

Barbour Compendium, Building Publishers Ltd., 1 Pemberton Row, London EC4P 4HL, UK

BIOSIS Information System, User Services Department, 2100 Arch Street, Philadelphia PA 19103, USA

BLAISE Online Services, The British Library, Bibliographic Services, 2 Sheraton Street, London W1V 4BH, UK

Bonn University, Abt. Zentralbibliothek der Landbauwissenschaft, Meckenheimer Allee 172, D–5300 Bonn 1, West Germany

British Library Document Supply Centre, Boston Spa, Wetherby, West Yorkshire LS23 7BQ, UK

British Library Science Reference & Information Service, SRIS, 25 Southampton Buildings, London WC2 1AW, UK

British Standards Institution, 2 Park Street, London W1A 2BS, UK

Building Research Establishment, Garston, Watford WD2 7JR, UK

Cambridge Scientific Abstracts, 7200 Wisconsin Avenue, Bethesda MD 20814, USA

Canada Institute for Scientifc and Technical Information, CISTI, Building M–55 Montreal Road, Ottawa K1A 0S2, Canada

Centre National de la Recherche Scientifique, CNRS, *see* Institut de l'Information Scientifique et Technique, INIST

Chadwyck-Healey Ltd., Cambridge Place, Cambridge CB2 1NR, UK

Cimtech *see* National Centre for Information Media and Technology

Cologne Zentralbibliothek der Medizin, Joseph-Stelzmann-Strasse, D–5000 Cologne, 41, West Germany

Commisariat à l'Energie Atomique, Centre d'Etudes Nucléaires, Saclay, France

Commission of the European Communities, Office of Official Publications, L–2985, Luxembourg

Commonwealth Agricultural Bureaux, CAB International, Wallingford OX10 8DE, UK

Commonwealth Scientific and Industrial Research Organisation, CSIRO, PO Box 225, Dickson ACT 2602, Australia

Council of Europe, Bôite Postale 431 R6, 67006, Strasbourg Cédex, France

Cranfield Institute of Technology, Cranfield, Bedford MK43 0AL, UK

Croner Publications Ltd., 173 Kingston Road, New Malden, Surrey KT3 3SS, UK

Cronos-Eurostat, I/S Datacentralen af 1959, Retortvej 8, DK–2500 Valby, Copenhagen, Denmark

DATA Inc. 9889 Willow Creek Road, PO Box 26875, San Diego CA 92126, USA

Defence Research Information Centre, DRIC, Kentigern House, 65 Brown Street, Glasgow G2 8EX, UK

Defense Technical Information Center, DTIC, Cameron Station, Alexandria VA 22304–6145, USA

Department of Energy, Thames House South, Millbank, London SW1P 4QJ, UK

Department of the Environment, 2 Marsham Street, London SW1P 3EB, UK

Department of Trade and Industry, 1–9 Victoria Street, London SW1H 0ET, UK

Department of Transport, 2 Marsham Street, London SW1P 3EB, UK

Derwent Publications Ltd., Rochdale House, 128 Theobalds Road, London WC1X 8RP, UK

Deutsche Forschungs- und Versuchsanstalt für Luft- und Raumfahrt eV, DFVLR, Linder Hohe, Postfach 90 60 58, D–5000 Cologne 90, West Germany

DIALOG Information Services Inc, 3460 Hillview Avenue, Palo Alto CA 94304, USA

Economic and Social Research Council, 160 Great Portland Street, London W1N 6BA, UK

Educational Resources Information Center, ERIC, Office of Educational Research and Improvement, US Department of Education, Washington DC 20208, USA

Environmental Chemicals Data and Information Network, ECDIN, Joint Research Centre, Ispra Establishment, 21020 Ispra (Varesa), Italy

European Parliament, Secretariat, Centre Europeen, Kirchberg, Luxembourg

European Space Agency, ESA, 8–10 rue Mario Nikis, 75738 Paris Cédex 15, France

Eurostat *see* Cronos-Eurostat

Fachinformationszentrum, FIZ, Energie Physik Mathematik GmbH, Zentral Fachbibliothek D–7514, Eggenstein-Leopoldshafen 2, West Germany

Food and Agriculture Organisation of the United Nations, FAO, Via delle Terme di Caracalla, 00100 Rome, Italy

Food and Drug Administration, Health and Human Services Department, 5600 Fishers Lane, Rockville, MD 20857, USA

Fund for Open Information and Accountability, FOIA, 145 W 4th Street, New York NY 10012, USA

Her Majesty's Stationery Office, HMSO, Publications Centre, PO Box 276, London SW8 5DT, UK

Information Publications International, IPI, White Swan House, Godstone, Surrey RH9 8LW, UK

Inkadat *see* Fachinformationszentrum, FIZ

INSPEC Information Services, Institution of Electrical Engineers, Station House, Nightingale Road, Hitchin, Herts SG5 1RJ, UK

Institut de l'Information Scientifique et Technique, INIST, Château du Monlet, 54514 Vandoeuvre-les-Nancy, Cédex, France

Institute for Scientific Information, ISI, European Branch, 132 High Street, Uxbridge UB8 1DP, UK

International Atomic Energy Agency, IAEA, PO Box 100, A–1400 Vienna, Austria

International Association of Agricultural Librarians and Documentalists, IAALD, c/o Centrum voor Landbouwpublikaties en Landbouwdocumentatie, PUDOC, Gen. Foulkesweg 19, Postbus 4, 6700 AA Wageningen, Netherlands

International Federation of Library Associations, IFLA – copies of documents on the IFLA International Programme for UAP are available from the British Library Document Supply Centre

International Food Information Service, IFIS, Lane End House, Shinfield, Reading RG2 9BB, UK

International Labour Organisation, ILO, Vincent House, Vincent Square, London SW1P 2NB, UK

International Register of Potentially Toxic Chemicals, IRPTC, United Nations Environment Programme, Palais des Nations, 1211 Geneva 10, Switzerland

International Standard Book Number, ISBN, *see* Standard Book Numbering Agency Ltd

International Translations Centre, ITC, Schuttersveld 2, 2611 WE Delft, Netherlands

Japan Information Centre of Science and Technology, JICST, CPO Box 1478, Tokyo, Japan

Kelly's Directories, Windsor Court, East Grinstead House, East Grinstead, West Sussex RH19 1XB, UK

Kompass Publishers, Windsor Court, East Grinstead House, East Grinstead, West Sussex RH19 1XD, UK

London Research Centre, (formerly GLC Library), Parliament House, Black Prince Road, London SE1 7SJ, UK

Microinfo Ltd., PO Box 3, Omega Park, Alton, Hants GU34 2PG, UK

National Aeronautics and Space Administration, NASA, Code NTT–4, Washington DC 20546, USA

National Agricultural Library, NAL, US Department of Agriculture, Beltsville, MD 20705, USA

National Centre for Information Media and Technology, Cimtech, Hatfield Polytechnic, College Lane, Hatfield, AL10 9AB, UK

National Foundation for Educational Research in England and Wales, NFER, Upton Park, Slough SL1 2DQ, UK

National Research Council of Canada, NRCC, Ottawa KLA OR6, Canada

National Micrographics Association *formerly* National Microfilm Association *see* Association for Information and Image Management, AIIM

National Technical Information Service, NTIS, US Department of Commerce, 5285 Port Royal Road, Springfield VA 22161, USA

National Translations Center, John Crerar Library, Chicago IL 60616, USA

Northern Ireland Council for Educational Research, The Queen's University of Belfast, 52 Malone Road, Belfast BT9 5BS, UK

Office National d'Etudes et de Recherches Aérospatiales, ONERA, 29 avenue de la Division LeClerc, 92320 Chatillon, France

Physik Verlag GmbH, Postfach 1260, D–6940 Weinheim, Germany

The Planning Exchange, 186 Bath Street, Glasgow G2 4HG, UK

RAND Corporation, 1700 Main Street, PO Box 2138, Santa Monica CA 90406–2138, USA

Royal Aircraft Establishment, RAE, Ministry of Defence, Farnborough, Hants GU14 6TD, UK

K G Saur Verlag, Postfach 71 10 09, D–8000 Munich 71, West Germany

Society of Automotive Engineers, SAE, European Office, American Technical Publishers Ltd., 68a Wilbury Way, Hitchin, Herts SG4 0TP, UK

Standard Book Numbering Agency Ltd., J. Whitaker & Sons Ltd., 12 Dyott Street, London WC1A 1DF, UK

System for Information on Grey Literature in Europe, SIGLE, British Library Document Supply Centre, Boston Spa, Wetherby, West Yorkshire LS23 7BQ, UK

Technical Indexes Ltd., Willoughby Road, Bracknell, Berks RG12 4DW, UK

Télésystèmes Questel, 83–85 boulevard Vincent-Auriol, 75013 Paris, France

Transport and Road Research Laboratory, TRRL, Department of Transport, Crowthorne, Berks RG11 6AU, UK

Transportation Research Board, National Research Council, 2101 Constitution Avenue NW, Washington DC 20418, USA

United Nations Educational Scientific and Cultural Organisation, Unesco, 7 place de Fontenoy, Paris 75700, France

United Nations Publications, C109, 1211 Geneva 10, Switzerland

United States Department of Energy, Office of Technical Information, PO Box 62, Oak Ridge TN 37381, USA

Universitatsbibliothek Hannover und Technische Informationsbibliothek, UB/TIB, Welfengarten 1B, D–3000 Hannover 1, West Germany

University Microfilms International, 300 North Zeeb Road, Ann Arbor, Michigan 48106, USA

Warwick University, Coventry CV4 7AL, UK

Index

United States Atomic Energy
  Commission, 13, 131
United States Government Printing
  Office, 28, 126
Universal Availability of Publications,
  34
Universal Decimal Classification, 42,
  93
University Microfilms International, 60
Usage and abusage, 50

Value of research and development
  reports, 156

Warwick University, working papers
  collection, 112–115
Watergate, 20
Weinberg report, 17
Whitaker's books in print, 80
White Paper on official secrets, 20
Working papers, 112
Worksheets, bibliographic control, 45
Writing technical reports, 50

York conference on grey literature, 6